WHY SCIENCE NEEDS ART

Why Science Needs Art explores the complex relationship between these seemingly polarised fields. Reflecting on a time when art and science were considered inseparable and symbiotic pursuits, the book discusses how they have historically informed and influenced each other before considering how public perception of the relationship between these disciplines has fundamentally changed.

Science and art have something very important in common: they both seek to reduce something infinitely complex to something simpler. Using examples from diverse areas including microscopy, brain injury, classical art and data visualisation, this book delves into the history of the intersection of these two disciplines before considering current tensions between the fields. The emerging field of neuroaesthetics and its attempts to scientifically understand what humans find beautiful is also explored, suggesting ways in which the relationship between art and science may return to a more co-operative state in the future.

Why Science Needs Art provides an essential insight into the relationship between art and science in an appealing and relevant way. Featuring colourful examples throughout, the book will be of interest to students and researchers of neuroaesthetics and visual perception, as well as all those wanting to discover more about the complex and exciting intersection of art and science.

Richard AP Roche is Senior Lecturer in the Department of Psychology, Maynooth University, Ireland.

Francesca R Farina is Postdoctoral Research Fellow at Trinity College Institute of Neuroscience, Trinity College, Dublin.

Seán Commins is Senior Lecturer in the Department of Psychology, Maynooth University, Ireland.

WHY SCIENCE NEEDS ART

From Historical to Modern Day Perspectives

Richard AP Roche
Francesca R Farina
Seán Commins

Routledge
Taylor & Francis Group
LONDON AND NEW YORK

First published 2018
by Routledge
2 Park Square, Milton Park, Abingdon, Oxon OX14 4RN

and by Routledge
711 Third Avenue, New York, NY 10017

Routledge is an imprint of the Taylor & Francis Group, an Informa business

British Library Cataloguing in Publication Data
A catalogue record for this book is available from the British Library

Library of Congress Cataloging in Publication Data
A catalog record for this book has been requested

ISBN: 978-1-138-95922-4 (hbk)
ISBN: 978-1-138-95923-1 (pbk)
ISBN: 978-1-315-66074-5 (ebk)

Typeset in Bembo
by Out of House Publishing

CONTENTS

ACKNOWLEDGEMENTS

We would like to offer our sincere thanks to Drs Noel Fitzpatrick and Tim Stott (Dublin Institute of Technology and GradCam) for their valuable input and feedback on early drafts of Chapter 1. We also gratefully acknowledge funding support from the Maynooth University Publication Fund 2017 for permissions and reproduction costs.

FOREWORD

Both artists and scientists by traditional definitions are interested in describing the universe and asking questions about its nature. Art and science's common root is a desire to communicate our ideas about the human experience. Our procreative drive underlies the dispersion of our genes, whereas our inherent desire to create and understand underlies the dispersion of our ideas, or epigenes. The principal distinction between science and art is from where the practitioner draws their inspiration – from the external (our understanding of 'objective' reality) vs internal (subjective reality) experience, respectively.

The most successful and interesting works of human ingenuity often integrate elements from both ends of this spectrum. Scientific understanding of the concrete rules that govern the universe is an immensely useful knowledge base from which to draw, enabling the combination of those ideas in ways unique to each person. An artist's palette is exponentially expanded when their knowledge of their materials and methods deepens through scientific understanding; likewise, a scientist is more creative and flexible when employing the orthogonal strategy of subjective thinking, the ability to dream up daring hypotheses for which there is yet no hard scientific evidence.

A thorough understanding of a mystery as daunting and complex as the brain requires as formidable an arsenal as possible with which to attack the topic. Why do we imagine that the only people among us making any real contributions to neuroscience are those who wear white lab coats and tirelessly toil over their instruments? What about the Impressionists, for example – the artists who began to experiment with visual stimuli by breaking them down into rough strokes for the first time, giving your mind mere impressions of form? These scientist–artists used their subjective reasoning to experiment with the degree of detail of form and colour that the viewer's brain needed in order to construct an image. They understood that your brain can infuse emotive qualities into non-photorealistic visual stimuli, a finding that has since been explored by anatomical and functional studies performed in the lab.

It is time for us to abandon the constraining and simplistic notion that a person can be either artist or scientist, never both. There is no such thing as a pure artist or a pure scientist. We all employ thinking derived from observations of both our objective and subjective realities. Employing every faculty, we know that honing the sharpest solutions should be the highest guiding principle of how we can most effectively grapple with the challenging problems that await humanity in the coming decades.

The current volume explores these issues in a wide-ranging discussion of the intersections of art and science, both historically and in the future.

Greg A Dunn,
Neuroscientist and Artist

1
THE INCOMPLETE MIND

Scientists and artists are not as different as they may seem. For most, the stereotypes associated with each are very familiar – the passionate, volatile artist in a chaotic studio; the cold, impassive scientist in a sterile, orderly laboratory. These caricatures are, in almost every sense, polar opposites of each other. Yet this is simply not the case. These cartoonish representations are easily refuted by reference to any of the plethora of passionate, tempestuous scientists or systematic, methodical artists that we can name. But the deeper claim – that the two are fundamentally opposed – is just as easily waved away by pointing to the numerous and profound similarities that artists and scientists share. For at their core, the approaches of art and science are intrinsically related, so much so that we may consider them siblings.

When we think of art, the concept that most easily comes to mind is consistent with the Western tradition of pictorial, figurative representations – painting, drawing, sculpture and other such visual forms. This is, of course, a narrow and culture-bound view, as art extends beyond the visual to encompass the verbal, and the line between these two can often be blurred: verbal art can be visual – poetry, metaphor, typography – just as easily as visual can be verbal. Verbal and visual art can also occupy the same space comfortably. Bearing this broader view in mind[1], we will continue to discuss mainly figurative forms of visual art and the creators thereof.

The commonalities between artists and scientists are many – they are both interested in reducing the infinitely complex to something simpler: the intricate structure of a tree to a few strokes of a brush; the relationship of matter and energy to a single, concise equation. They both strive to express that which is difficult to comprehend in a purer, more elegant way. Both must observe the world very closely. A large portion of the artist's task is to simply look, study and note things about the environment around them[2]: colours, shapes, light and shadow, texture, detail and motion. All of these aspects must be closely regarded and understood before pencil is put to paper, brush to canvas, chisel to stone. Likewise, the scientist must scrutinise phenomena very minutely, sometimes over and over again to validate their observations. Notes must be taken, anomalies recorded, patterns discerned and changes monitored. In both cases, artists and scientists spend much of their time looking before they act.

Artists and scientists employ specific methods in order to pursue their goal. Whether these techniques involve the manipulation of paint on a palette or variables in an experiment, the moulding of clay or the handling of data, they are skills which must be learned and perfected over time before work can commence. This period of apprenticeship is often long – historically, years of learning in the presence of a master; years of study in the laboratory of a supervisor. In both cases, these skills and techniques are taught, acquired, refined and honed through repeated use, feedback, development and eventual expertise. And yet, even then, for both the artist and the scientist, there is always more to be learned. To paraphrase the great astronomer and science communicator Carl Sagan, science and art are ways of thinking much more than they are bodies of knowledge.

The work of both artist and scientist can have a deep and profound effect on people. This may come via a shocking or evocative piece of visual art or through a new and controversial scientific theory. The statements that they make, whether verbally in hypotheses or visually through a painting, sculpture or installation, can often elicit the strongest feelings in those that witness them. And, in both cases, these feelings can be overwhelmingly positive – elation, awe, excitement – or negative – hate, suspicion, prejudice or simply bewilderment (for example, modern art or scientific jargon). Regardless, it can be said that few topics are more divisive than art and science.

One further fundamental similarity that artists and scientists share is what they endeavour to achieve through their work. Both are concerned, at the deepest level, with the expression of fundamental truths in some form: insights about ourselves, about the world around us, about the nature of nature itself. Both seek to enlighten, illuminate and enable people to better comprehend themselves, their world and their relationship to it. In short, they attempt to promote a deeper understanding of everything.

In recent years, we have seen the emergence of a new field of study, neuroaesthetics, in which the methods of neuroscience – such as brain imaging – have been applied to the study of what humans find beautiful, aesthetically pleasing or artistically interesting. This scientific endeavour to reveal the neurobiological underpinnings of our sense of vision, visuality[3] and aesthetic appreciation has been greeted by some with deep scepticism and suspicion. Such objectors claim that the attempt to isolate and quantify the brain areas and neural architectures which allow us to appreciate great art somehow diminishes the achievement of the artist or removes an element of the mystique from the creative process. Here we will consider some of the early findings from this fledgling field and outline how some principles of perception and visual information processing can help us to better appreciate the astonishing skills and techniques used by artists to create their desired effects. In so doing, we will allow the reader to decide for themselves whether this knowledge detracts from the experience of appreciating these works or enhances it.

Other critics[4] claim that while art has learned much from science, science, in turn, can extract nothing new or valuable from art. This view comes in response to the increasing number of recent collaborations between artists and scientists on projects which seek to find a synthesis of the two areas, with generous funding for such projects awarded by the UK's Wellcome Trust, among others. Happily, some scientific images have begun, in recent years, to be appreciated for their aesthetic merit as well as their relevance to research, leading some to ask why these images from science should not be considered 'Art' with a capital 'A'[5]. In many branches of science, prizes are now awarded for the most striking or beautiful images obtained from experimental data, and an academic journal has been founded – the *Art and Science Journal* – to celebrate images which bridge the divide between these two spheres. Yet the question posed by critics of this approach is a pertinent and important one: what – if anything – has art given to science?

This book seeks to discuss these ideas by tracing the history of the relationship between art and science, considering current tensions between the two disciplines, and looking to the future of how these two seemingly polarised approaches to understanding may be able to co-exist under the rubric of neuroaesthetics. The book begins by recounting the Renaissance, a time when art and science were seen as inseparable and symbiotic pursuits, an approach best epitomised by that colossus of both fields, Leonardo da Vinci, who wrote:

> Principles for the Development of a Complete Mind: Study the science of art. Study the art of science. Develop your senses – especially learn how to see. Realize that everything connects to everything else.
> *Attributed*

We then describe how, even after these two paths to knowledge have diverged, there remain many examples of art influencing and performing an important role for science; these examples are taken from diverse areas such as microscopy, brain injury and data visualisation. Finally, we turn our attention to the future, and how the emerging field of neuroaesthetics may enable the relationship between the two to return to a more co-operative, interdependent state by using science to discover how certain pieces of art are so affecting and uplifting.

Throughout this book we will provide examples of how aspects of art – and, more specifically, visual forms of art – have historically given something to science. We will present images which have helped to steer the direction of scientific fields of inquiry, made a complex idea more comprehensible, or revealed something unexpected about an aspect of the human condition. Such images convey stories of where art and science converge and intersect, encircle each other and occupy the same space at the same time. They represent the reciprocal relationship of these two routes towards self-knowledge along which we navigate – of art underpinning science and science enriching art. They can be considered signposts to connections between these parallel roads. Roads which once were one, and which are converging once more.

Definitions and remit

This volume will deal with issues involving science, art and aesthetics; it is instructive, therefore, to begin with some definitions, to aid clarity and to make explicit the specific remit of this volume in terms of what is and is not included within this discussion.

In relation to science, this is relatively straightforward – science can be defined in several ways: as a collection of knowledge or facts, derived from observations and experiments, giving rise to laws or principles; as a system for gleaning such facts, a set of rules and procedures which allow us to explore difficult questions and arrive at answers; as a way of thinking, characterised by critical thought and a need for observable and reproducible demonstrations of phenomena before an idea can be accepted; as a self-informing and self-revising system of knowledge wherein no question is forbidden or considered unaskable. Science – in its ideal form – is a way of doing things free of subjectivity, bias, agendas and presuppositions, a pure approach to uncovering knowledge and understanding built on the solid foundations of reductionism, empiricism, positivism and objectivity.

Some of the finest scientific minds of the past two centuries have defined it in various eloquent ways. Biochemist and science fiction writer Isaac Asimov describes science as a way of testing one's ideas against the cosmos[6]. Astronomer and advocate of women in science Vera Rubin takes this idea further, stating that science gives us the freedom to challenge our preconceptions[7] and to learn that not all teachings are true[8]. Above all, science reminds us that there are always new discoveries to be made. In the words of the pioneering chemist Marie Curie:

> One never notices what has been done; one can only see what remains to be done.
>
> *Letter to her brother (1894)*

When it comes to defining art and aesthetics, though, things become trickier. 'Art' is notoriously difficult to define, and its long association with the idea of beauty has only served to muddy the waters[9], while even the term itself requires further clarification as to whether Fine Art is being referred to or Applied Art/Information Design, which may include Visual Culture, Cognitive Art(s), Visualisation/Visuality or Visual Aesthetics. In this regard, we acknowledge that our remit here will not be all-encompassing; we will largely focus on Western examples and definitions of *visual aesthetics and visual representational art* – outlined below – in the subsequent chapters. In this section, we summarise the main definitions of and approaches to art and aesthetics, describing – in a general and broad overview – the key movements and schools of thought in these areas in order to clarify what aspects of these broad topics fall within the remit of the current volume.

Definitions and theories of art

What constitutes 'Art' and what qualifies something as a 'work of art' are difficult philosophical questions about which volumes have been written; we will not seek to answer these questions here. For many centuries, dating back to the great thinkers of Ancient Egypt, Greece and Rome, the idea that art equated to beauty (specifically beauty in nature) was influential, leading to the eventual emergence of aesthetics as an independent field of study in 1750. Such notions would later be scathingly (and gleefully) eviscerated by Tolstoy in *What is Art?* (1898). What Tolstoy does provide, however, is a useful definition of the *purpose* of art, which in his view is directed at communicating feeling and emotions – the ugly, fearful and obscene as well as the beautiful, sublime and transcendent – to the viewer by means of perception. In this he proposes an important and complementary role for science – he views the role of science as being to explain and communicate *knowledge*, while art seeks to express *feelings*, with both ultimately employed in the task of improving the lot of humankind.

> Science and art are as closely bound together as the lungs and the heart, so that if one organ is perverted the other cannot act rightly.
>
> *Tolstoy (1962, p. 210)*

In this conception of the symbiotic roles of art and science, Tolstoy touches on the crucial role of visual representational art in science communication, an idea which we explore in detail in Chapters 2–4.

Aside from definitions of art or its purpose, several intellectual approaches to art have arisen, each of which take a slightly different approach to art's subject matter and nature. The Historical approach, for example, focuses primarily on the idea that, to fully appreciate a work of art, one must possess considerable knowledge of the skills and techniques required to produce it and also of the specific socio-cultural, political or historical circumstances surrounding its creation. We must remember that many – or most – works of art were not originally intended to adorn the walls of museums; rather, they were created for specific patrons and places or for the artists themselves. In such an approach, the background to a work's creation is as important as the artistic intention of its creator, and the absence of either of these will detract from the aesthetic experience of the viewer[10].

The Psychological approach, by contrast, concerns itself with the idea that the presence of some quality – most frequently beauty – defines a work as 'Art'; in this way it adheres to the idea of *universalism*, which proposes that a concept such as beauty is an independent quality present or inherent in certain objects or images, rather than a response to a stimulus within the perceiver. This universalist idea has led some to attempt to seek out a locus for the brain's response to such a quality (considered more in Chapter 8 on neuroaesthetics).

Finally, Expressionist theories[11] hold that the aim of art is to convey emotions that are typically difficult to express verbally; therefore, unlike the Historical and Psychological/Universalist approaches, the Expressionist view does not wed art closely to the concept of beauty. From this approach also comes the idea of art engendering a feeling of 'disinterested interest', a reaction whereby a person is engaged by a work without the desire to acquire, control or manipulate it, which can be considered comparable to the difference between 'liking' and 'wanting'.

Definitions and theories of aesthetics

Due to the long-standing historical tendency to conflate the ideas of art and beauty, the concept of aesthetics is frequently and often inextricably linked to the definition of art. Aesthetics – which was defined by neuroscientist Anjan Chatterjee in 2011 as 'the perception, production and response to art, as well as interactions with objects and scenes that evoke an intense feeling, often pleasure' – was first coined by Alexander Baumgarten in 1750, the term itself derived from the Greek for 'perception'. The philosopher Immanuel Kant (1724–1804) proposed in 1790 that the discipline of aesthetics should focus on the concept of beauty in art and nature, and while he did not agree with the idea that art equated to beauty, he was a universalist in that he believed that beauty was a quality which could be found in both nature and art[12]. In the nineteenth century, this aesthetic viewpoint became synonymous with the Philosophy of Art, arguably due to the conflation of Kant's questions about the sensible experience with questions about aesthetic judgement (e.g. genius, the sublime, beauty).

Within Philosophical aesthetics there are two broadly-defined schools – Continental and Analytic aesthetics. Continental aesthetics, which originated with Kant, tends to focus on issues such as thought, language, and social and political factors in a view of art as something that is lived and performed in a cultural context. It draws upon ethical, political and ontological considerations, emphasising sensory experience, embodiment, culture and metaphysics in the context of art and aesthetics. By contrast, Analytic aesthetics, which has been prominent since the 1950s, tackles questions about art by using logical arguments to consider clear and precise statements about art and aesthetics. It involves the *ahistorical* analysis of concepts, sometimes informed by empirical observations, in the study of the aesthetic qualities of objects (including those in nature, perhaps suggesting an element of universalism). In recent decades, Analytic Aesthetics has begun to devote increasing attention to individual arts, including music, literature, painting, dance and others.

Empirical aesthetics, on the other hand, attempts to use scientific approaches (such as experimentation and case studies) to study the nature of the aesthetic experience. Under this rubric, Descriptive aesthetics seeks to understand and explain aesthetic responses in terms of known psychological and physiological phenomena, while Experimental aesthetics – which originated with pioneering psychophysicist Gustav Fechner (1801–1887) – invokes the methods of experimental psychology in order to study aesthetic experiences. Such approaches have given rise to theories of the processing stages of aesthetic experience like the *Aesthetic Triad*, wherein three stages – sensory–motor, emotion–evaluative and meaning – are engaged when viewing an aesthetic piece, or the *12 Components of Aesthetic Experience* proposed by Pelowski and colleagues (in 2016). The structure of Empirical aesthetics has, in recent years, been effectively mapped onto the emerging field of *neuroaesthetics*, which we discuss towards the end of this volume.

Professor Peter Weibel[13], CEO of the Center for Art and Media, Karlsruhe, and Dr Ljiljana Fruk, lecturer in bionanotechnology at the University of Cambridge, have proposed that the 'Art System' (or artistic establishment) has been far too narrow in what it construes as art. They trace the history of art from Classical Antiquity through the Renaissance, Modernism and Postmodernism, noting the shift in emphasis of the content of art – from image to reality, from the visual *representation* of objects to the *presentation* of objects, from depicting the surface (the visible) to revealing the interior (the invisible) – thereby moving from painting, drawing and sculpture to the camera, technology and media art. They urge that recent forms of technology-assisted media art (e.g. by microscopes, telescopes and x-rays) that gave rise to images of cells, viruses, planets and others – all of which Francis Bacon would term 'Res Invisibles' – *should* be considered art. This is because many such pieces remain true to the original agenda for aesthetics proposed by the philosopher Alexander Gottlieb Baumgarten, namely the investigation of sensory cognition. Weibel and Fruk see this endpoint as a natural extension of the trajectory originally set in motion by Leonardo da Vinci, in his *Trattato della pittura*, whereby the principles of aesthetics progressed from point to line to surface to the body enclosed by that surface. In fact, Leonardo was among the first to delve beneath the surface in the course of his anatomical investigations (discussed more in Chapter 2), correctly demonstrating that a superior understanding of the invisible structures below the skin would yield more accurate representations of the surface features. Weibel and Fruk underscore the way many advances in science, such as those listed above, have facilitated this evolution of art towards its inevitable endpoint, highlighting the importance of a dialogue between art and science.

> Going forward, we will need a science of the image that includes handmade images by artists, and technical images made by artists as well as scientists. In the future, art will be weak visualisation, and science will be strong visualisation.
>
> *Weibel and Fruk (2013, p. 63)*

Having considered these (somewhat over-general) distinctions between art and aesthetics, we can at this point define precisely how we intend to use the term 'art' throughout this volume. For the purposes of this discussion on the interactions and interdependences of art and science, we will use the term 'art' to refer to *visual representational art or aesthetics*. In some places, we will alternate the term 'art' for 'aesthetics' for the purposes of variety, but we intend to use these terms interchangeably. And while it can be claimed that this represents a somewhat narrow definition of art which can be considered a largely white, Western, male, nineteenth-century view of art, we openly acknowledge this bias on our part and make no claims to the contrary; further, given the examples we intend to discuss, this selectivity would seem warranted[14]. Furthermore, since proponents of the view that art should play no role in science primarily rest their arguments on similar definitions of art, it is necessary to adopt such a definition in order to evaluate their claims appropriately.

What this book is about

In this volume, we will explore the shared histories of art and science, considering the overlapping roles of art and science during the Renaissance (Chapter 2) and our understanding of such diverse topics as brain cells and celestial bodies (Chapter 3) to more recent examples of how visual art has promoted the development of data visualisation (Chapter 4). In the second part of the book, we consider the neuroanatomical apparatus which underpins the human visual system, citing the remarkable story of an artist as a metaphor for its operation (Chapter 6). We then describe a number of cases in which artworks have revealed something important about the function of the brain and where brain injury has had a profound effect on the nature of artworks produced by the injured person (Chapter 7). Finally, in Chapter 8 we discuss neuroaesthetics, the attempt to isolate the brain regions implicated in the act of aesthetic appreciation or artistic experience, and we return to the arguments against a role for art informing science, evaluating this claim in the light of the preceding examples. Just as Weibel and Fruk argue that art has benefited – in ways not yet fully recognised by the artistic community – from developments in science, we make the complementary argument that art has, historically, informed and advanced the content of science in many ways and can continue to do so. We propose that a continued dialogue between art and science constitutes a bidirectional enhancement which will be to the benefit of both.

Notes

1 It is noteworthy that the famous Dutch painter Piet Mondrian was one of many who tried to reduce the Western tradition of figurative painting to universal abstract principles (such as the perpendicular relation of two lines) or to unmodulated fields of primary colours.
2 Although observation of the world is no longer a necessary or sufficient condition of art in the twenty-first century.
3 See Baxandall (1994) for more on visuality.
4 Most notably scientist and author Professor Lewis Wolpert.
5 In his exceptional book, *A Story of Art*, EH Gombrich (1950) discusses the concept of what many of us call Fine Art, and from where this idea stems.
6 Interview by Bill Moyers (1988, pp. 5–6).
7 As quoted in Koupelis and Kuhn (2007, p. 583).
8 An idea also discussed by the physicist Richard Feynman (1969).
9 The pros and cons of this conflation are nicely discussed in Tolstoy (1898).
10 One of the defining theories of modern art history is that vision is historically contingent. As Heinrich Wölfflin (1915) put it: 'Not everything is possible at all times. Vision itself has its history, and the revelation of these visual strata must be regarded as the primary task of art history.'
11 Such as those proposed by Anthony Ashley Cooper, the Third Earl of Shaftesbury (1671–1713).
12 In Kant's view, aesthetics should be considered the original project of the science of the sensible experience, thereby enabling a distinction between beauty and art.
13 Weibel and Fruk (2013).
14 'Why Have There Been No Great Women Artists?', written in 1971 by Professor Linda Nochlin, provides an excellent discussion of the factors that have traditionally precluded women and minority groups from achieving the status of 'genius' in art.

References

Baxandall, M. 1994. Fixation and Distraction: The Nail in Braque's Violin and Pitcher (1910). In Onians, J. (ed.), *Sight and Insight: Essays for E. H. Gombrich*. London: Phaidon, pp. 399–415.

Feynman, R. P. 1969. What Is Science. *The Physics Teacher*. 7(6), pp. 313–320.

Gombrich, E.H. 1950. *The Story of Art*. London: Phaidon.

Koupelis, T. and Kuhn, K.F. 2007. *In Quest of the Universe*. Burlington: Jones & Bartlett Learning.

Moyers, B. D. 1989. *A World of Ideas: Conversations with Thoughtful Men and Women about American Life Today and the Ideas Shaping Our Future (Vol. 1)*. New York: Doubleday Books.

Nochlin, L. 2003. Why Have There Been No Great Women Artists? In Jones, A. (ed.), *The Feminism and Visual Culture Reader*, pp. 229–233. London: Routledge.

Tolstoy, L. 1962. "*What is Art?" and Essays on Art*, trans. Aylmer Maude. New York: Hespirides.
Weibel, P. and Fruk, L. (Eds.). 2013. *Molecular Aesthetics*. Cambridge, MA: MIT Press.
Wölfflin, H. 1950. *Principles of Art History: The Problem of the Development of Style in Later Art*. New York: Dover.

2

ART AND SCIENCE AS ONE

Introduction

For over 500 years, Leonardo da Vinci has been the subject of unending study. His assorted accomplishments – in the fields of art and architecture, engineering, anatomy, and science – have been commemorated in thousands of text books, and his appearance and personality (even his sexuality) have been imagined and debated. He is regarded as the archetypal 'Renaissance Man', a polymath and a visionary thinker – in short, a genius. As Sigmund Freud once remarked, Leonardo was a man who woke too early, at a time when everyone else was still sleeping[1]. Given this impressive and long-standing reputation, it would be fair to wonder what – if anything – is left to say about Leonardo da Vinci that has not already been said[2]. For this reason, the present chapter will not include a biography of Leonardo, nor will it chronicle his many accolades. Instead, our aim is to illustrate how Leonardo da Vinci was, in many ways, as much a scientist as he was an artist. Through his story, we will demonstrate that the principles of scientific and artistic endeavour are not as different as we may think.

The idea that art and science could be intimately connected is, perhaps, a strange one to us today, but this would not have been the case in the world as Leonardo knew it. He was born in 1452 in the town of Vinci, near Florence, at the very heart of the Renaissance movement – a period of artistic and literary revolution beginning in Italy and subsequently spreading throughout Europe. At the time, the word '*art*' was primarily used to denote a 'skill' or 'technique' and the word '*science*' signified 'knowledge'. Within Italian society, artists and medical workers were represented by a single merchant group – the

Guild of Physicians and Apothecaries (along with surgeons, undertakers, wine distillers, booksellers and silk traders) – and painters commonly purchased their pigments in the same shops where doctors bought medicines. If we imagine for a moment what it must have been like to live in this environment, we can begin to understand the reasoning behind Leonardo's proverbial 'note-to-self' quoted in Chapter 1: 'realise that everything connects to everything else'.

Leonardo was an exemplary student in this regard, studying painting, sculpture, music (he was an accomplished lyrist), architecture, engineering, anatomy, astronomy, geology, apothecary and botany over the course of his lifetime. Chief among these disciplines was, of course, art. Despite never finishing most of his works, Leonardo is celebrated as one of the greatest artists of all time. The Mona Lisa is often used as a prime example of his artistic abilities; in this painting, Leonardo succeeds in captivating the viewer using just a single enigmatic feature: a smile[3]. His talent as an artist, however, was not merely the result of natural ability (which he undoubtedly possessed in abundance). It was refined and improved by dedicated study and methodical practice. To Leonardo, painting was a science[4], and so it required a scientific method.

The artist's scientific method

Within the broad field of science, investigations of the natural world are carried out using a series of standardised practices which have come to be known as the 'Scientific Method'. While the exact order differs somewhat across disciplines, the Scientific Method typically includes the following steps: (1) make an observation about the world, (2) construct a theory that attempts to explain this observation, (3) perform an experiment to test your theory, (4) examine the results of the experiment and draw conclusions, and (5) repeat the process, refining each iteration as you go. Although the Scientific Method as we know it today was not formalised until around the seventeenth century[5], Leonardo was (as usual) ahead of his time. In fact, in 1568, the Italian painter Giorgio Vasari (1511–1574), friend and biographer of Leonardo, wrote that 'he might have been a scientist if he had not been so versatile'.[6] But it was precisely his versatility that enabled Leonardo to apply his own Scientific Method to his artwork. This simple method comprised three steps: observation, experimentation and repetition.

Observation

In his Divine Comedy, the poet Dante Alighieri[7] reminds us that 'art, as far as it is able, follows nature, as a pupil imitates his master'. Leonardo, in turn, was a dedicated pupil of art. Time and again in his notebooks he discusses the importance of learning 'how to see' and that Nature itself is the best book from which to learn. There are numerous anecdotes of Leonardo's sharp eye. In one such story, Vasari describes Leonardo's habit of following people whom he found striking or unusual, sometimes for hours at a time. He would then return home, where he would draw them perfectly from memory. There are also stories of how Leonardo would invite people from the lower social classes to dine with him and tell them fantastical tales so that he could observe natural delight without the restraints of 'good breeding'. He is even thought to have spent time with convicted criminals in their final hours so that he could witness true sorrow and despair. Such was his grasp of emotion and expression, he became adept at caricatures (or 'grotesques'), in which he skilfully exaggerated what he saw (see Figure 2.1). Drawing was undoubtedly Leonardo's preferred method of dissemination. He believed that visual representation was superior to description:

> Dispel from your mind the thought that an understanding of the human body in every aspect of its structure can be given in words; for the more thoroughly you describe, the more you will confuse the mind of the reader and the more you will prevent him from a

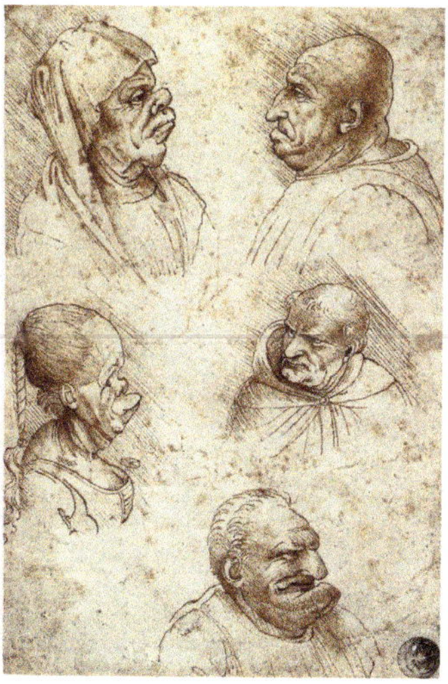

FIGURE 2.1 Left: *Study of Five Grotesque Heads* (1494), Leonardo da Vinci, [Public domain], via Wikimedia Commons. Right: *Five Caricature Heads* (1490), Leonardo da Vinci, [Public domain], via Wikimedia Commons.

> knowledge of the thing described; it is therefore necessary to draw as well as describe … I advise you not to trouble with words unless you are speaking to a blind man.
>
> *Playfair McMurrich (1930)*

Not satisfied with 'learning to see', Leonardo endeavoured to understand the mechanisms of *how* we see. He was fascinated by the senses (particularly the eye and optic nerve; Figure 2.2) and perception. One drawing from 1508 depicts one of Leonardo's theories, wherein he hypothesised that the optics of the eye could be enhanced by placing the cornea in direct contact with water, either by placing one's head in a glass bowl filled with water and looking through the bottom or by wearing a water-filled structure over

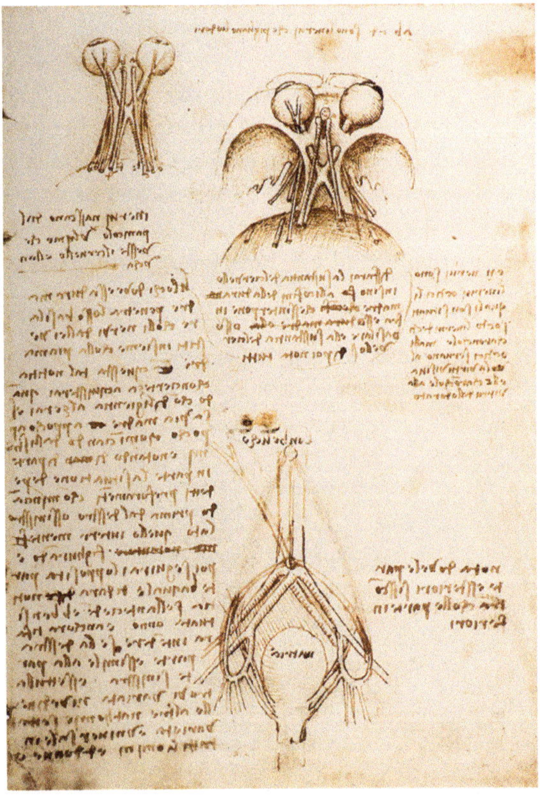

FIGURE 2.2 Detail from *The Optic Chiasma and the Cranial Nerves* (1506), Leonardo da Vinci, reproduced with permission from Dennis Hallinan/Alamy Stock Photo.

FIGURE 2.3 *Study for the Adoration of the Magi* (circa 1481), Leonardo da Vinci [Public domain], via Wikimedia Commons.

the eye. This theory is regarded as the first known conceptualisation of the contact lens.

Leonardo also understood the 'rules' of seeing. He was a master of light and perspective. This can be seen plainly in many of his preparatory drawings (such as Figure 2.3). His masterpiece, *The Last Supper*, also illustrates these principles beautifully (see Figure 2.4). The realistic sense of depth is seamlessly achieved through linear perspective (converging behind Christ's head) and aerial perspective (seen in the landscape outside the windows).

Theory and experimentation

Like any good scientist, Leonardo did not rely on observation alone. Instead, he valued hands-on practice. In his mind, those who relied solely on authority or doctrine were servants to memory and not their own intellect. He would frequently defend this idea against the academics of the time in his notebooks, stating that experience – and not memory – was mother to all sciences and arts:

> Though I may not, like them, be able to quote other authors, I shall rely on that which is much greater and more worthy: on experience, the mistress of their masters.
>
> *Richter (1888, p. 11)*

FIGURE 2.4 *The Last Supper* (1498), Leonardo da Vinci (online) [CC BY-SA 4.0 (http://creativecommons.org/licenses/by-sa/4.0)], via Wikimedia Commons.

Leonardo's enthusiasm for acquiring practical experience was shaped by his apprenticeship with Andrea del Verrocchio (1435–1488) in Florence circa 1470. Verrocchio – whose name translates to 'true eye' – fervently believed that in order to become an accomplished artist, one must first fully understand the subject of one's drawing. For example, when drawing a nude figure, one must first know the underlying musculature of the human body and the mechanics of how it operates. To achieve this, Verrocchio encouraged the practice of 'anatomies', or dissections. Human dissection had been revived in Italy in the eleventh century due to a growing interest in understanding illness and disease (and to investigate the occasional suspected poisoning). As such, they were a well-established – often public – practice throughout Italy by the time the young Leonardo arrived at Verrocchio's studio. Sometime in the 1480s, Leonardo began carrying out his own dissections. He would go on to dissect more than 20 human cadavers of all ages. In his notebooks, he describes his thoughts on this practice:

> And if you have a love for such things, you will perhaps be hindered by your stomach, and if this does not prevent you, you may perhaps be deterred by the fear of living during the night in the company of quartered and flayed corpses, horrible to see. If this does not deter you, perhaps you lack the good draughtsmanship which appertains to such demonstrations, and if you have the draughtsmanship, it will not be accompanied by a knowledge of perspective. If it were so accompanied,

> you lack the methods of geometrical demonstration and the method of calculation of the forces and power of the muscles. Perhaps you lack the patience so that you will not be diligent. Whether all these qualities were found in me or not, the hundred and 20 books composed by me will supply the verdict, yes or no.
>
> *da Vinci (1952)*

Though admittedly gruesome, Leonardo emphasises how important it is for a painter to be a good anatomist. In looking at his drawings (Figure 2.5), it is easy to see the benefit of his anatomical study. Without this practical knowledge, he states, the artist risks their figures looking like a 'bag of nuts' or a 'bundle of radishes'. Unfortunately for Leonardo, some of his contemporaries disagreed with his methods and labelled him a heretic. Soon after, his privileged access to cadavers was revoked by Pope Leo X, which permanently halted his anatomies.

Over the course of his anatomical studies, Leonardo produced many detailed drawings of the human brain and skull from varying perspectives, showing cranial nerves and meningeal arteries (arteries connected to the membranes which enclose the brain) (see Figure 2.6). He also wrote a short guide for removing a brain and devised a novel method for studying the ventricles (cavities within the brain) by injecting hot wax into an ox brain to produce a cast, thereby demonstrating his ingenuity as an artist and a scientist. He was also the first person to correctly identify that the heart has four chambers, not two, and that adult males and females have the same number of teeth. Again, he emphasises the importance of visual representation in his experimentation, stating that 'all science that ends in words has death rather than life'.

In his later years, Leonardo began to appreciate the value of anatomies beyond a purely aesthetic viewpoint. One illustrative example is his dissection of a man known as 'the centenarian', with whom Leonardo spent time before his death. It is reported that the man's passing was so peaceful that Leonardo endeavoured to find the cause of a death 'so sweet'. In this case, the artist's goal was not artistic; rather, he sought to understand the scientific process of ageing and the secret to long life. It is also worth noting here that Leonardo valued repetition of methods. Likewise in science, it is considered best practice to repeat one's experiments several times before settling on any one conclusion. This ensures that any result is not likely to be due to error or fluke, but instead represents a genuine discovery. Vasari reports that whenever possible, Leonardo would dissect a region twice (in two different individuals) so that he could compare and note the differences, thereby minimising any error. In spite of these efforts, Leonardo made many errors over the course of his studies. For example, he often confused human and animal anatomy

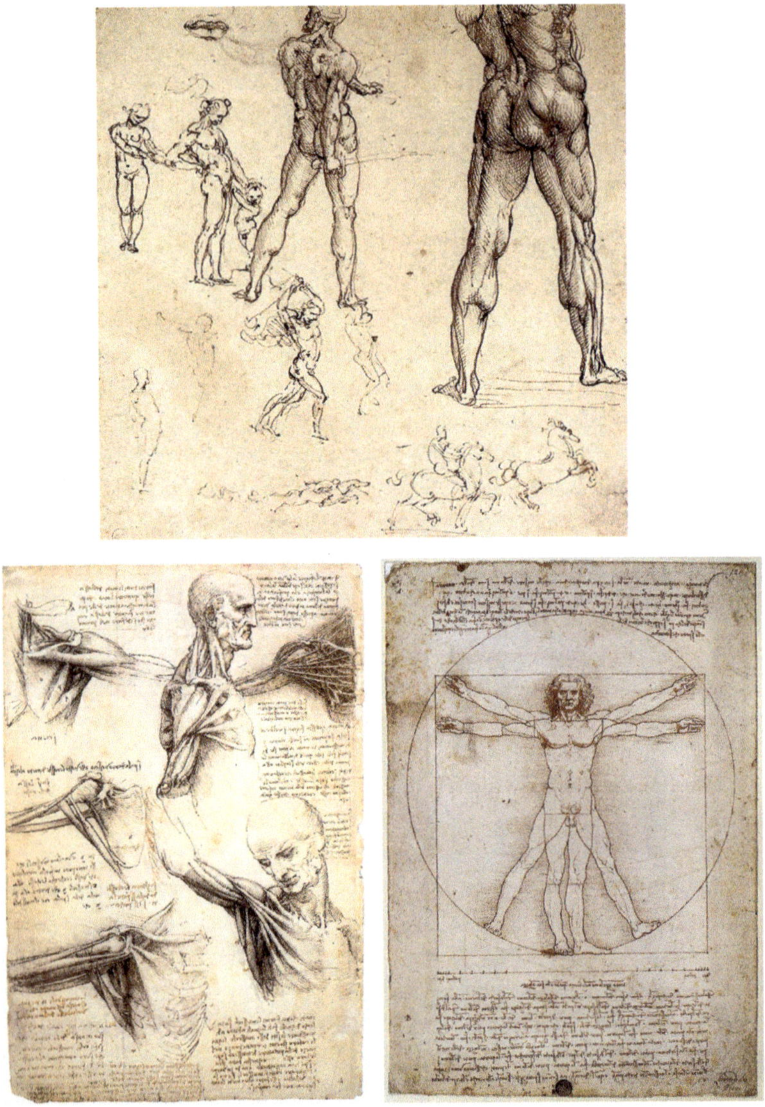

FIGURE 2.5 Top: *Anatomical Studies* (1505). Bottom left: *Anatomical Studies of The Shoulder* (1510). Bottom right: *The Vitruvian Man* (1487). All by Leonardo da Vinci [Public domain], via Wikimedia Commons.

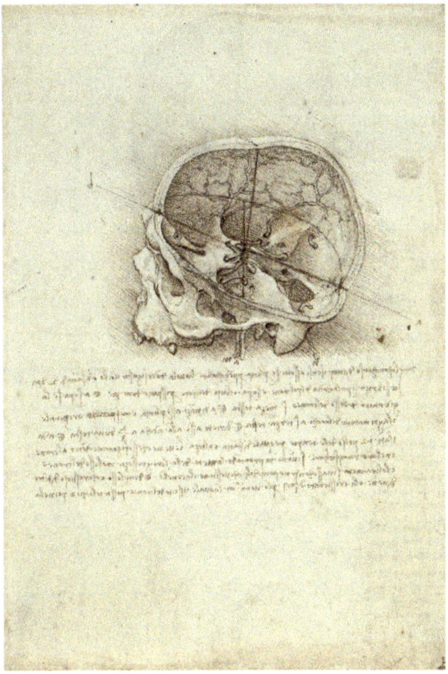

FIGURE 2.6 *View of a Skull* (circa 1489), Leonardo da Vinci [Public domain], via Wikimedia Commons.

(in keeping with the inherited knowledge of the time). However, as the late scientist Thomas M Schofield said, 'science is not about finding the truth at all, but about finding better ways of being wrong'[8].

Art and science as one

After nearly 20 years of study and experimentation, Leonardo's intention was to produce a comprehensive treatise on anatomy, entitled *De Figura Humana*, by the year 1510. His notes include detailed descriptions of how the book should be laid out, beginning with illustrations of the womb and foetus, followed by the infant, child, and finally the adult female and male[9]. He even boasted that these drawings would be of such quality that the viewer would feel almost as if they had a real person of flesh and bone presented before them. Sadly, Leonardo died before his book was completed and so his anatomical studies were not published for another 300 years. Although it is difficult to say whether he would ever have produced his anatomical text, even if had lived another twenty years. It is well known that Leonardo was prone to new fascinations, and roughly two-thirds of all his artistic pieces remained unfinished. It is even rumoured that on his deathbed Leonardo expressed regret over having offended God with the quality (and quantity) of his work.

What we can conclude is that Leonardo set the scene for those who came after him in the fields of art and science. This includes his famous rival Michelangelo Buonarroti (1475–1564). Michelangelo was twenty-three years younger, and so would have grown up studying Leonardo's illustrations and methods. Leonardo's infamous reputation would have been an obvious source of motivation for Michelangelo (as well as countless other artists).

Leonardo is also thought to have been a source of inspiration for Renaissance anatomists. Andreas Vesalius (1514–1564) was a Flemish anatomist and physician who is today regarded as the father of modern anatomy. In 1543, Vesalius published a collection of anatomical books entitled *De Humani Corporis Fabrica*, which represented a significant advancement in anatomical knowledge at the time. As part of this collection, Vesalius included a series of engravings to accompany the text. These illustrations show human skeletons standing in animated poses, thinking, some even mourning their own death (Figure 2.7). In addition

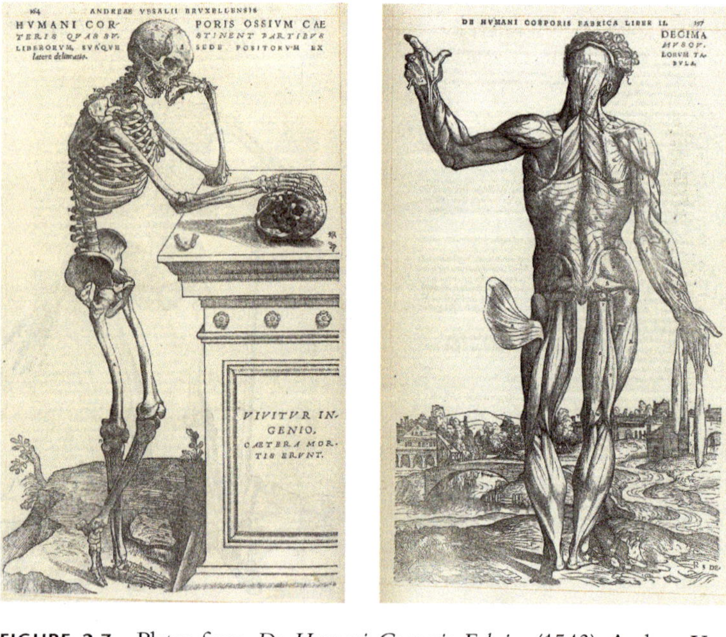

FIGURE 2.7 Plates from *De Humani Corporis Fabrica* (1543), Andreas Vesalius, [Public domain], via Wikimedia Commons.

to lively postures, some corpses are given delicate curled hairstyles and others are presented in front of detailed landscapes. The beauty of these drawings drew wide acclaim, leading some to suggest that Vesalius may have seen Leonardo's original drawings for his Treatise and stolen the designs for himself[10]. However, it seems unlikely that Vesalius would have been aware of Leonardo's intention to publish an anatomy text, and thus of the drawings' existence – though it is possible. Regardless of their origin, Vesalius' engravings contributed to the far-reaching impact of his books because they represented more than a mere description; they gave life to his words. Like Leonardo before him, Vesalius understood the value of illustration as a method of dissemination, and chose to represent science and aesthetics in harmony.

Notes

1 Freud (1916).
2 An internet search for 'Leonardo da Vinci' returns nearly 9 million unique results as of January 2018, and large portions of his personal papers ('The Codex Arundel') have recently been digitised by the British Library. It seems there is no shortage of information about this great thinker for those who are interested.
3 Gombrich (1950, chapter 15) discusses the merits of the Mona Lisa in detail.
4 Pevsner (2002) discusses this further.
5 Due in large part to the works of the Italian astronomer Galileo Galilei (1564–1642) and English philosopher Francis Bacon (1561–1626).
6 Vasari (1987, p. 187).
7 Alighieri (2017, Canto XI. 103).
8 Schofield (2013).
9 Described in Playfair McMurrich (1930, pp. 74–75).
10 Playfair McMurrich (1906).

References

Alighieri, D. 2017. *Inferno*. In *The Divine Comedy*, Canto XI. Aegitas.

da Vinci, L. 1888. *The Notebooks of Leonardo da Vinci*, trans. J.P. Richter. Project Gutenberg, www.gutenberg.org/etext/5000.

da Vinci, L. 1952. *On the Human Body*, trans. Charles D. O'Malley and J.B. de C.M. Saunders. New York: Schuman.

Freud, S. 1916. *Leonardo da Vinci*. New York: Moffat, Yard & Co.

Gombrich, E.H. 1950. Harmony Attained. In Gombrich, E.H., *The Story of Art*. London: Phaidon.

Pevsner, J. 2002. Leonardo da Vinci's Contributions to Neuroscience. *Trends in Neurosciences*. 25(4), pp. 217–220.

Playfair McMurrich, J. 1906. Leonardo da Vinci and Vesalius: A Review. *Medical Library and Historical Journal*. 4(4), pp. 338–350.
Playfair McMurrich, J. 1930. *Leonardo da Vinci: The Anatomist (1452–1519)*. Washington, DC: Carnegie.
Schofield, T.M. 2013. On My Way to Being a Scientist. *Nature*. 497, pp. 277–278.
Vasari, G. 1987. *Lives of the Artists: Volume 1*, trans. George Bull. New York: Penguin.

3

SEEING FURTHER, SEEING SMALLER

Introduction

With Leonardo da Vinci, we saw how his artistic genius was fundamental to his ability to depict and disseminate what he observed in relation to human anatomy and mechanics. Yet, brilliant as anatomists like da Vinci and Vesalius were, their insights were restricted by the limitations of what they could observe. In this chapter, we will consider how artistic representations were crucial for scientists who wished to convey things which could not be seen with the naked eye. For Galileo Galilei and Santiago Ramón y Cajal, being able to accurately reproduce what they saw through their respective eyepieces – whether the stars in the night sky or the neurons of the brain – was a crucial tool in reporting their findings. But while Galileo and Cajal intentionally employed art as a means to relate scientific facts, there are other examples – including Giotto, van Gogh and the Bayeux Tapestry – where events of scientific import were captured almost accidentally in artistic works. Finally, the converse situation will be considered as we discuss artworks which have been claimed to conceal, rather than reveal, scientific observations. These examples further highlight the relationship between art and science at a time when few other means existed to communicate complex ideas to a wider audience.

Galileo: through a lens, brightly

Carl Sagan famously said, 'Somewhere, something incredible is waiting to be known.' As is often the case in science, truths – when they are finally revealed – do not arrive all at once; they often show themselves in a gradual, piecemeal fashion so that slowly, incrementally, an accurate picture builds up. While da Vinci had concerned himself with learning 'how to see', it was Galileo Galilei (1564–1642) who showed the world something that could not be seen. By pointing his newly-invented telescope towards the heavens, Galileo would (eventually) bring about a radical expansion of thinking and redefine our understanding of the known universe. His insights would not have been possible without advances in technology in the seventeenth century which afforded possibilities that were not present before.

Although his name is now synonymous with the instrument of his discoveries, Galileo did not invent the telescope himself. Perhaps unsurprisingly, the original idea can be attributed to da Vinci in the fifteenth century; nestled in his notes within the *Codex Atlanticus*, Leonardo wrote 'Construct glasses to see the moon magnified' (p. 168), although there is no evidence that he ever constructed one. It would take a German-born spectacle maker working in Holland, Hans Lippershey, to attempt to put the concept into practice. In 1608 Lippershey applied for a patent for a refracting telescope with three-times magnification, which he described as a device 'for seeing things far away as if they were nearby'. His application, however, was unsuccessful, with the Dutch government ruling that the invention was too easy to replicate. In fact, by the summer of 1609, the Englishman Thomas Harriot had improved the design and constructed a functioning telescope with six-times magnification. It was around this time that Galileo was introduced to the instrument while visiting friends in Venice. Within months, he had fashioned his own spyglass (or *perspicillum*), his early attempts made from lead piping with lenses at either end. Through successive iterations and improvements to the design, Galileo arrived at a telescope which yielded an image nine times larger than that possible with the naked eye. Directing this instrument toward the moon would not allow him to see the entire surface all at once, just one-quarter at a time – but at a greater resolution than had ever been seen before. With the aid of this device, the celestial bodies – like other truths – would give up their secrets little by little.

Galileo Galilei was born on the 15th of February 1564 in Pisa, a city which was then under the Grand Duchy of Florence. His father, Vincenzo Galilei, was a lute player and nobleman of an ancient family line. The young Galileo was provided with a good education, showing an early interest in and aptitude for his studies, particularly mathematics. At the age of eight, the family moved to Florence, by which time he was beginning to demonstrate signs of his creativity and ingenuity. He liked to make toys and models, enjoyed painting and was a good musician. Discouraged from pursuing art by his father, Galileo was pushed towards medicine, although he would not complete his medical degree but instead returned to his first love, mathematics. His fascination with mathematics was born of his interest in drawing and music, and his desire to learn the underlying principles of each, while his inquisitive approach and disregard for received wisdom was sometimes a source of frustration among his professors. He was appointed professor of mathematics in Pisa in 1589, aged 26, moving to the University of Padua two years later to teach geometry, mechanics and astronomy until 1610. It was during this period, in the winter of 1609, that Galileo first raised his spyglass skyward. Between November and December, he documented the six phases of the moon, and from these initial examinations he arrived at a simple – but radical – conclusion:

> I feel sure that the surface of the Moon is not perfectly smooth, free from inequalities and exactly spherical, as a large school of philosophers considers with regard to the Moon and the other heavenly bodies, but that, on the contrary, it is full of inequalities, uneven, full of hollows and protuberances, just like the surface of the Earth itself, which is varied everywhere by lofty mountains and deep valleys.
>
> *Galileo (1610, p. 7)*

This controversial statement was in direct opposition to the traditional views of the moon held by both Aristotle and the Catholic Church. They believed that the moon and planets were considered to be perfectly smooth examples of absolute sphericity, a testament to God's ability to create perfection. The impeccable smoothness of the moon was linked to the purity of the Virgin Mary according to the Church, and older explanations of any observed bumps or spots on the moon claimed that its surface was like a mirror, reflecting the imperfections of the Earth.

As with the invention of the telescope, Galileo was not the first person to view the moon under magnification – Harriot had done so from London four months earlier, using a 'perspective tube' of six-times magnifying power. However, Harriot's observations were accompanied by one only crude drawing, and he made no attempt to explain what he had seen. Galileo, by contrast, complemented his descriptions with a series of seven watercolour paintings of what he had perceived (one for each moon phase and a seventh from the following January; see Figure 3.1). Each painting was exquisite in its realism, showing evidence of techniques favoured by Renaissance artists such as *chiaroscuro* (the contrasting of light and shade); these images are all the more remarkable given that they were produced from memory. The paintings were

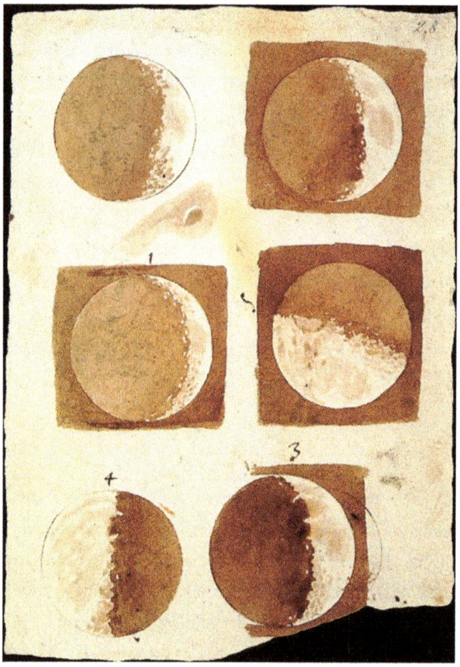

FIGURE 3.1 *Drawings of the Moon, November–December 1609*, Galileo Galilei [Public domain], via Wikimedia Commons.

reproduced as engravings when Galileo published his observations in the text *Siderius Nuncius* ('The Starry Messenger', 1610), allowing the world – for the first time – to see the moon in its true form. Galileo's understanding of artistic principles such as perspective and contrast allowed him to produce images which (for the most part) accurately depicted what he saw *and* allowed him to explain what they meant, effectively allowing the public to peer through his telescope. People could now look, see and judge for themselves, rather than relying on doctrine and faith for their understanding of the heavens and the bodies that occupied them. The publication of the *Siderius Nuncius* would constitute the first blow in an assault on the Church's view of the cosmos on a scale it had never encountered. It was the first breeze of a sea-change in thinking.

With his spyglass, Galileo continued to hypothesise and investigate, studying the surface of the moon and successfully calculating the height of the mountains he observed. Using the fourth version of his telescope – which afforded him thirty-times magnification – he turned his attention to other objects in the night sky, gazing deeper into the darkness and discovering Io, Europa, Ganymede and Callisto, the four largest moons of Jupiter. With characteristic humility (or, some would argue, political astuteness), rather than naming them after himself, Galileo called them 'Cosimo's Stars' after his patron, the Grand Duke of Tuscany, Cosimo de Medici II (and later the 'Medician Stars' in honour of the Medici family); they would inevitably be renamed the Galilean Moons after their discoverer. In 1610, he was rewarded for his discoveries with an appointment as Court Mathematician/Philosopher to Cosimo II – a promotion which was also probably due in no small part to his celebrity status after improving the telescope, and its military applications[1]. During this period, he depicted newly observed stars in the Milky Way (specifically along Orion's Belt) which could not be seen by the unassisted eye, leading him to conclude that these stars did not get their light from the sun, as had previously been assumed. These discoveries appeared together when he published his seminal *Siderius Nuncius,* concluding this volume with the promise of more discoveries to follow; Galileo knew there was yet more to see.

Despite the controversial and inflammatory nature of his scientific observations and the opposition which they inevitably elicited, Galileo did enjoy some support from his contemporaries in the art world. Among these supporters was lifelong friend and artist Lodovico Cigoli (known as *Cardi*). The two had known each other from their time with Ostilio Ricci, who taught them both geometry and perspective in Pisa. As his final artistic work, Cardi (1559–1613) painted a homage to Galileo's scientific discovery in his fresco in the Santa Maria Maggiore in 1612; the *Assunta* depicted exactly how Galileo had described the moon in the *Nuncius*, its surface uneven and pockmarked. Sadly, in a grim foreshadowing of things to come, the fresco would be painted over four years later. Cardi maintained that Galileo 'could see better' because he was better prepared by his artistic training and knew how to draw. In his view, a man lacking the ability to draw was a man who lacked eyes.

Galileo's artistic prowess was such that he even attained membership of the art academy of Florence (*Accademia del Designo*) – the prestigious academy established by Giorgio Vasari, artist and biographer of da Vinci – four years after his watercolours of the moon. It is debatable as to whether he might have found comparable renown as an artist rather than a scientist, had he opted for a different path when the fork in the road was before him. Viviani (his biographer and final student) recounts that Galileo busied himself 'with great delight and marvellous success in the art of drawing, in which he had such great genius and talent that he would later tell his friends that if he had possessed the power of choosing his own profession at that age … he would absolutely have chosen painting.' Yet it appears that his talents were reserved mainly for mathematical drawing – although he did doodle; the original version of the *Nuncius* contains some charming drawings of buildings, a boat and a river. What is clear, however, is that he could bring his artistic talents to bear on the scientific messages he sought to convey, employing them in service of his message and ensuring his observations reached a wider audience and were met with greater understanding than they might, had it been otherwise.

His supporters in the art world aside, with each subsequent discovery Galileo's ideas were met with increasingly staunch opposition – from both the Church and the public at large; a public who refused to, or were not ready to, believe. In July of 1610 he observed the rings of Saturn, despite not knowing what they were (referring to the effect as 'a planet with ears'); an explanation would not arrive until 40 years later with Huygens. He recorded the phases of Venus in December of that year and documented sunspots for the first time in 1611. Unfortunately, however, the growing unease about the implications of his ideas culminated in him being banned by the Inquisition from talking about his views in 1616 (the same year that Copernicus' controversial text *On the Revolutions of the Heavenly Spheres* was banned).

Although he accepted this ban, he nevertheless returned to his work of investigating the skies and in 1632 produced a second text, *Dialogue Concerning the Two Chief World Systems*. This volume tells the fictional story of three men – two philosophers and a layman. Salviati (named after a deceased friend) and Simplicio were the philosophers, the former being a supporter of Galileo's findings concerning heliocentrism and the latter being a supporter of traditional (Church-supported) geocentric views. In the story, the layman, Sagredo (named after another friend), questions the philosophers over the course of four days of discussions. Through this discussion, Galileo not only asserted that the moon's surface was uneven, but also extolled the idea of heliocentrism in place of geocentrism, as beautifully described in the following line:

> The sun, with all those planets revolving around it and dependent on it, can still ripen a bunch of grapes as if it had nothing else in the universe to do.
>
> *(1632)*

The heliocentric view, originally advocated by Copernicus in 1543, directly opposed the accepted wisdom at the time that the Earth was the centre of the universe – fixed, immovable and special. All planets, the sun, the stars, every celestial body was positioned at varying distances from the central Earth. Galileo, being acutely aware of the potential of this incendiary view to draw more of the Church's ire upon himself, sought the approval of the Vatican for his *Dialogue*. However, an outbreak of plague in Florence meant that communication with Rome was impossible; as a result, the book was approved locally without the Church's consent. When the text was eventually reviewed in Rome, it failed to meet the approval of the Vatican, as Pope Urban VIII (some claim mistakenly) interpreted the character name Simplicio as a device to satirically imply that the Church was stupid. An order for the recall of the text was issued, but by that time a thousand copies had already been distributed, to the anger of the Church. And so, Galileo was brought back before the Roman Inquisition in 1633 (aged 70) and this time labelled a heretic. He was tortured, forced to recant his views[2] and sentenced to prison, though this was later commuted to house arrest for the remainder of his life. He lived out this period in his villa in Arcetri, in the southern hills of Florence. By this time he was in poor health, his sight failing, his reputation damaged and his life's work outlawed. The severity of his treatment at the hands of the religious establishment also had a wide-reaching influence on others: on learning of Galileo's sentence, René Descartes suspended publication of his book on heliocentrism, further suppressing the idea. In late 1641, the ailing scientist, now fully blind and suffering from arthritis, contracted a fever from which he would never recover. It is perhaps one of history's bitterest ironies that the man who gave – and endured – so much to make the world see, a world that remained wilfully blind for so long, should reach the end of his life without his sight.

Galileo Galilei died on the 8th of January 1642, at the age of 77. As a final act of punishment from the Church, he was denied the right to be interred next to his ancestors within the main body of the Basilica di Santa Croce in Florence, and despite protests from the Grand Duke of Tuscany Ferdinando II (son of Galileo's patron Cosimo II) no monument was erected in his honour. Instead, his body was placed in an unmarked grave and relegated to a remote corner of the church where it remained for close to 100 years. Galileo was justly relocated to the main area of the Basilica in 1737, but it would be another century before the Church would rescind their ban on his works, and it was only in 1992, some 350 years after his death, that Pope John Paul II issued a formal apology to Galileo. Not unlike his Florentine predecessor, Leonardo da Vinci, Galileo's findings faced an arduous path towards dissemination and eventual acceptance. But today, through the lens of hindsight, he is rightfully recognised as the Father of Modern Science for his contribution to an intellectual revolution, one that would have a profound effect on our understanding of the cosmos and our place in it. As Sagredo tells Salviati in the *Dialogue Concerning the Two Chief World Systems*, all truths are easy to understand once they have been discovered, 'the point is in being able to discover them'.

Cajal's *Butterflies of the Soul*

Galileo had pointed his telescope toward the heavens and peered at the sky, producing drawings which would – eventually – lead to a change in the way we viewed our place in the solar system. Over 200 years later another European, Santiago Ramón y Cajal (1852–1934) of Spain, would look through a similar configuration of lenses, this time directing his gaze inward, not at the stars but at a target almost as great in number: the neurons of our brain. Like Galileo, the illustrations he created of the images under his eyepiece would result in a dramatic shift in thinking within his branch of science. But unlike Galileo, the route to acceptance of these new ideas would be considerably less fraught.

Santiago Ramón y Cajal was born in 1852 in the village of Petilla de Aragon in the Navarre region of Spain, under the shadow of the Pyrenees mountains. Although he would go on to follow in his father's footsteps as a student of medicine and anatomy, the man who became known as the Father of Neuroscience planned initially to become an artist, a career choice indicative of his rebellious nature. Faced with early discouragement from this idea, just like Galileo before him he embarked upon his medical studies at Zaragoza, where his father also held an appointment in the anatomy department. It was during the dissections he carried out at this time that the young Cajal found an outlet for his artistic talents, producing intricate and exquisite drawings of the tissue samples he examined. He would stare for long periods of time through the microscope at the tissue before him, observing, noting, counting, before turning to a blank page to reproduce from memory the complex structures and physiological arrangements he had seen. This habit of creating illustrations which were both accurate and beautiful would become a feature of Cajal's work throughout his career, and would pave the way for a revolution in the way we think about the workings of the brain.

Cajal's great insight would come about due to a confluence of two crucial elements: an instrument and a technique. As Galileo had his telescope through which the nature of the solar system was revealed to him, so too did Cajal have an instrument by which his discoveries were made possible – the microscope.[3]

Microscopy had become commonly used across Europe from the 1600s onwards; the term 'microscope' itself was coined by Giovanni Faber in 1625 for, appropriately, Galileo's compound microscope – so by the late nineteenth century the technology had advanced considerably, providing excellent resolution at high levels of magnification. When Cajal settled in Valencia in the early 1880s, with a faculty position in the university and a growing family there, he continued his interest in examining and reproducing with detailed sketches the anatomical structure of all manner of samples, including tissue obtained from local butcher shops and morgues. But through all his investigations, he was restricted by the limitations of the techniques available to him – one of the primary problems with biological tissue is that it is extremely soft. To observe samples under the microscope, tissue must first be hardened in order to slice it, then preserved and finally stained with a chemical agent to render various structures visible. The fixation techniques Cajal was using were poor, and the staining agents (typically vegetable and aniline dyes) were inadequate for revealing the minutest detail. It would take the discovery of a new staining technique, by the Italian Camillo Golgi, for the mysteries of the brain's anatomical structures to reveal themselves to Cajal.

Golgi (1843–1926) had pursued his medical studies at Pavia, working alongside Giulio Bizzozero, a brilliant experimentalist and pioneer of the use of microscopy for medical research. Fascinated by the workings of the nervous system from an early stage, Golgi had dedicated his research to studying the billions of neurons that were visible in slices of brain tissue. At that time, little was known about the nature of these cells or how they were connected to each other, if at all. Such questions could not be answered given the techniques of the day; the chemicals used stained thousands upon thousands of neurons at a time. In the resultant forest of cells and fibres, it was impossible to tell where one neuron began and another ended. Confronted with this dilemma, Golgi experimented with hundreds of stains, systematically varying the timing and concentrations of the chemicals he used. Finally, in 1872, he discovered a method (involving the fixation of brain tissue in potassium dichromate and its subsequent immersion in silver nitrate) by which only a small percentage of the brain cells appeared stained in black. This allowed Golgi, for the first time, to identify and differentiate individual neurons among the web of interconnected cells. He could now discern the entire structure of the neuron, from the bulbous cell body with the dense array of branches (dendrites) protruding from it, down the extent of the long, tubular trunk (the axon) to the root-like terminal endfeet. Like Cajal, Golgi reproduced what he observed in the form of intricate and detailed drawings of brain structures, from the black dots of the cell bodies to the thin threads and filaments of the dendrites and axons. He termed the stain *la reazione nera* ('the black reaction'), and it would be this technique that would change the direction of Cajal's career and our understanding of the brain.

Although Golgi had discovered his black reaction in 1872, it was fifteen years before it would come to Cajal's attention. The relative isolation of Spain's scientific community meant that the technique was not widely known or used by scientists there, and it took a chance meeting in 1887 for Cajal to be exposed to the possibilities it presented. Having moved to Barcelona in this year, Cajal encountered the psychiatrist Luis Simarro Lacabra, recently returned from Paris with brain sections stained using the Golgi method. Upon seeing the tissue stained with the black reaction, Cajal's insatiable intellectual curiosity was awakened as the implications of this new technique became apparent to him. He later recounted this moment of epiphany in his autobiography as follows:

> Coloured brownish black even to their finest branchlets, standing out with unsurpassable clarity upon a transparent yellow colour. All was sharp as a sketch with Chinese ink.
>
> *Cajal (1923, p. 306)*

Until this point in his career, Cajal had been somewhat eclectic in the tissue samples he had studied and the topics he had addressed, ranging from bone to skin to other organs, and from cholera to inflammation. But from the moment he observed the stained slices of Lacabra, he devoted himself to the study of neural tissue using the Golgi method.

> Like the entomologist in pursuit of brightly coloured butterflies, my attention hunted, in the flower garden of the gray matter, cells with delicate and elegant forms, the mysterious butterflies of the soul, the beating of whose wings may someday – who knows – clarify the secrets of mental life.
>
> *(1923, p. 363)*

He set about applying the new staining technique to samples from different parts of the brain including the cerebellum ('little brain'), the spinal cord and the seahorse-shaped hippocampus. The microanatomy of each structure in turn laid bare before his exacting gaze, their secrets depicted – as ever – in the painstakingly intricate accuracy of his drawings. But more than merely describing the physical structure of these cells, as Golgi had done, for Cajal the images appearing under his lens and, later, on his page seemed also to provide him with hints as to their very function. He was looking at what Golgi had looked at, but seeing what he had failed to see. Summing up his excitement with typically evocative imagery, he stated:

> As new facts appeared in my preparations, ideas boiled up and jostled each other in my mind.
>
> *(1923, p. 325)*

Cajal studied many different types of neurons, from tree-like Purkinje cells to small basket cells to climbing fibres. He scrutinised their individual structures, from the crystalline tangle of dendritic branches to the trunk-like axon and the terminals so reminiscent of roots. As he peered more closely at these terminal regions of cerebellar cells under greater levels of magnification, he began to notice that the axons of one cell did not touch the dendritic tree of the next – there was a tiny, barely perceptible gap between the neurons. This was a ground-breaking finding; until this point, the dominant theory of the organisation of brain cells proposed that the billions of neurons were physically connected to each other with information (in the form of electro-chemical impulses) flowing in all directions. The implication that the neurons were effectively arranged in a mesh- or net-like configuration gave this doctrine its name: the Reticular Theory (*reticulum* coming from the Latin for net). Golgi himself was a proponent of this view, and he noticed nothing in his slices to dissuade him from it.

But Cajal's finding that neurons were, in fact, physically separated from their neighbouring cells led directly to the emergence of a new conceptualisation of cortical organisation, a theory which would become known as the Neuron Doctrine. His speculations on cellular function (based on structure) also supported this view; he surmised that the dendrites and cell body (soma) were the input regions of the cell, receiving electrical impulses from nearby neurons and passing this signal down the length of the axon to the terminal regions to be passed to the next cell in the circuit (now known as the law of functional polarity).

This shift in thinking from the Reticular Theory to the Neuron Doctrine represented a profound and important advancement in the way the brain was viewed. When Galileo had presented his revolutionary ideas two centuries earlier, his observations were met with threats, violence, claims of heresy and house-arrest. Fortunately for Cajal, the influential thinkers of the late nineteenth century were more open to novel proposals and to questioning the dominant theory of the day; as good scientists do, they were willing to abandon one theory in favour of another which better explained the data presented, and so the Neuron Doctrine became widely accepted[4]. But not by Golgi – he remained an advocate of the Reticular Theory, even after he was jointly awarded the Nobel Prize in Physiology or Medicine (shared with his bitter rival Cajal) in 1906 for his role in the work which directly discredited this view.

Cajal's discoveries owed much to his experimental rigour and restless intellect, but his drawings played a hugely important role in the dissemination and eventual acceptance of his ideas (see Figure 3.2). These drawings were never mere representations of what he viewed through the eyepiece of his microscope. They were excruciatingly accurate, down to the exact spatial arrangements of cells and the precise configuration of the dendrites and axons he observed. Further, he frequently created a sense of three-dimensional perspective in his sketches, with one neuron positioned in front of another to create the impression of depth, an effect achieved by bringing together

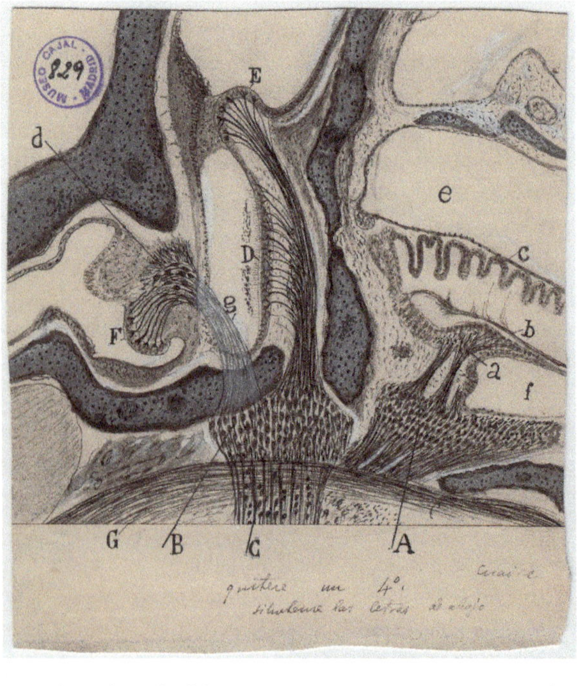

FIGURE 3.2 *The Labyrinth of the Inner Ear*, Santiago Ramón y Cajal, reproduced with permission from Instituto Cajal del Consjo Superior de Investigaciones Científicas, Madrid, © 2017 CSIC.

many drawings from multiple sections of a particular region. He strove to ensure that the image on the page bore as close a resemblance as possible to that seen through the lens, often 'touching up' his drawing with white gouache, removing axons/dendrites on a figure until he was completely satisfied with the likeness. Like Galileo, Cajal knew that drawing what he saw was the only way to adequately convey the content of his observations. He sought to create depictions of such clarity and detail that they could substitute for the act of microscopy itself, that to look at his drawings should yield the same image as to look through his lens. Without such rigour and dedication to his illustrations, it is difficult to know how much longer it would have been before the true nature of the brain's microstructure came fully into focus.

Caught on canvas: when art accidentally reveals science

The stories related above, of Galileo's celestial bodies and Cajal's neurons, demonstrate beautifully how, like the anatomists of the Renaissance before them, visual art could provide a telling contribution to the science of the day. And while they are comparable for their aesthetic merit as much as their contributions to knowledge, these examples share another feature – they were all conducted with the explicit intention to depict, elucidate and disseminate the observations of the scientist in question. But another subset of artistic works exists, one which also contributed to our understanding, but not through any intention on the part of the artist. These are cases where scientific features were captured by accident, caught in the background, as it were, of a snapshot of history.

The most famous example of such serendipitous insight is also the most promiscuous in that it appears in many separate pieces: Halley's Comet. Its first depiction was in the Bayeux Tapestry (embroidered in the 1080s), where its appearance of 1066 is represented, much to the fear and astonishment of onlookers. Arguably the most famous appearance of the comet, however, comes in the guise of the Star of Bethlehem in the fresco *Life of Christ: 2. Adoration of the Magi* by Giotto di Bondone (1266/7–1337). Here Giotto, inspired by the 1301 appearance of the comet, incorporates its fiery passage across the heavens into the biblical scene he painted in the years immediately afterward[5].

The comet would continue to make appearances in art on many of its 75-year return visits, notably in a drawing by the artist Diego Durán, whose depiction of the comet in Mexico circa 1516 represents an omen for the impending Spanish conquest. While many subsequent examples exist wherein Halley's Comet is, itself, the subject matter of the art – painting, jewellery, pottery and other media – its background appearances in the works of Giotto and others epitomise the occasional accidental contribution of visual art to the sciences[6]. Perhaps its next visit, due in July 2061, will contribute further.

The heavens also feature in another case of astronomy making an appearance in the works of a famous painter. Vincent van Gogh's (1853–1890) *Starry Night* famously includes an accurate representation of the constellations as they would have appeared in the night sky above the south of France in June of 1889, the time of its painting. According to the art historian Albert Boime, the positions of Venus, the triangle of Aries and the star Hamal are all in the locations they would have occupied at that time over Saint-Rémy-de-Provence. The only deviation from reality – the inclusion of a crescent rather than the almost full, gibbous moon – may be attributable to artistic preference: there is evidence that it was originally painted as a crescent moon and painted over. Nor was *Starry Night* unique in van Gogh's repertoire – astronomically accurate skies are also found in his other works *Café Terrace by Night* and *Starry Night over the Rhone River*. What's more, it has been suggested that the dynamic swirls and spirals of the celestial bodies in *Starry Night* may have been influenced by images of spiral nebulae or even, like Giotto before him, the transit of a comet.

There is also evidence for the presence of a more terrestrial (although no less dramatic) phenomenon in the works of Edvard Munch and Joseph MW Turner: volcanic eruptions. And while Giotto, van Gogh and others were at least conscious of including the contents of the night sky in their artworks, in the case of the after-effects of a volcanic event, the artists were probably unaware of its impact on their paintings. It is now thought that in the aftermath of a volcano's eruption, minute particles of ash, dust and gas – collectively referred to as aerosols – diffuse into the atmosphere. While the effects of these particles are normally unseen, for a narrow window of fifteen minutes after sunset the dusk sky is illuminated by vivid hues of orange, pink or purple by virtue of their presence, and these effects can be observed enormous distances away. Evidence suggests that such spectacular sunsets have repeatedly been captured by artists over the centuries, and that analysis of the ratio of red to green in the colours of the post-sunset sky can reveal the presence of such volcanic afterglows[7]. The eruption of Krakatoa in 1893 is said to have influenced the vivid skies in Edvard Munch's (1863–1944) *The Scream*, while a number of paintings by Joseph Turner (1775–1851) are also thought to have been guided by these atmospheric anomalies. This unusual situation demonstrates how scientific approaches – in this case the analysis of colour ratios in paintings – employed years later may reveal the tell-tale signs of an event of scientific import, even if the artists themselves were oblivious to them at the time. It appears that nature, as well as the artist, applied a signature to these pieces.

Paintings of human subjects may also reveal clues, often unintentionally, about different medical conditions; such illustrations sometimes provide accurate depictions of illnesses that might not have been well-recognised or understood at the time. For example, the work of Jusepe de Ribera (1591–1652) shows, in exquisite detail, specific anatomical abnormalities present in his models. Although often disturbing, there is a great realism in some of his work including *The Bearded Woman*, *The Clubfoot* and *The Grotesque Heads of Men*. These paintings are thought to reveal the fact that his subjects were suffering from, respectively, an endocrinological disorder, arthrogryposis and a thyroid goitre. Other works by de Ribera attempt to depict the suffering of early Christian saints and martyrs using models now thought to be afflicted with pectus excavatum. *Saint Onophrius* (1642), a fourth-century hermit who lived in the desert with little to eat or drink, is depicted as an old man with hanging flesh, hollow cheeks and chest deformities, a hallmark of pectus excavatum. Whether de Ribera was aware of the disorder – a condition earlier described by da Vinci[8] – is a matter for debate, but given the aspiration of many Renaissance artists to accurately represent what they observed, knowledge of the body's anatomy and of various conditions was fundamental to their work[9].

Such portraits provide modern scholars with a time- and location-stamp of the occurrence of particular disorders, as well as the social strata and approximate age in which they were commonly found. This analysis also extends to one of the greatest artists of the Renaissance, Michelangelo Buonarroti. Although many physical conditions have been linked to the great master – including lead poisoning, tophus arthritis and gout – recent analysis[10] of the portraits he produced shows evidence of a loss of dexterity experienced in his later life, probably due to osteoarthritis. Further, in Jacopino del Conte's (1510–1598) *Portrait of Michelangelo Buonarroti*, an ageing Michelangelo is depicted with his left hand hanging, showing signs of swellings at the base and other points along the thumb. These are not symptoms of gout but rather a form of osteoarthritis, probably brought about by decades of hammering and sculpting – the work leaving an enduring mark upon the artist.

Hidden anatomy: art concealing science?

The works considered so far in this chapter all have in common the fact that they managed – intentionally or otherwise – to *reveal* aspects of science through visual art. There are also cases of the converse situation, wherein scientific information is thought to be *concealed* within the content of artistic creations. Though highly speculative, the best-known example of this proposal involves one of the most famous works of art of the Western world, the Sistine ceiling.

In 1508, Michelangelo (1475–1564) was commissioned by Pope Julius II to repaint the vault (ceiling) of the Sistine chapel in Vatican City, Rome. Working from the entrance to the altar, Michelangelo chose to depict scenes from the Book of Genesis in a series of panels, including the *Creation of Adam* (above the entrance; started in winter 1511), the *Separation of Land and Waters*, the *Creation of the Sun and Moon* (over the altar; completed in summer 1512) and the *Separation of Light from Darkness*. These scenes depict God and either Adam or the various elements at the early stages of the biblical story of creation, and were the final four frescoes of the entire ceiling, completed in time for the reopening of the chapel in November of 1512. Several historians[11] have claimed that contained within these four panels are hidden anatomical images which highlight Michelangelo's mastery of human anatomy in addition to his prowess as an artist and architect.

The most well-known of these suggested anatomical hints appears in the panel *Creation of Adam*, in which God – resting on an amorphous drapery – reaches out to touch Adam, bestowing life upon him. Commentators have argued that the shape of the cloth surrounding God closely matches the outline of a human brain viewed from a sagittal (side-on) aspect, with key features such as the medulla oblongata, the optic chiasm and several sulci and gyri visible. They propose that the artist's intention was to suggest that God was conferring the gift of intelligence, as well as life, on Adam.

Further examples of hidden anatomy – again depicting the brain – are said to be visible in the final panel, the *Separation of Light from Darkness*, wherein God, viewed from below, stretches out his arms to divide these natural elements. Some have claimed that anomalies in the depiction of God's neck and inconsistent lighting patterns suggest that what Michelangelo painted here was not an anatomically inaccurate and technically inept view of the throat, but rather a physiologically realistic rendering of the structures of the brain stem as viewed – consistent with the viewer's vantage point – from below, with key structures (pons, medulla, spinal cord) all discernible. It is suggested that the anomalous lighting of this region of the figure is intended to draw attention to this area, perhaps providing an additional hint to its double meaning.

Despite being highly speculative, these and other claims – such as theories of additional brain structures hidden in the folds of God's tunic, for example – have received some support (e.g. Meshberger, 1990). Much is made of Michelangelo's friendships with anatomists such as Matteo Realdo Colombo (the Professor of Anatomy and Surgery who succeeded Vesalius at Padova), and the likelihood of his being familiar with da Vinci's anatomical illustrations, but solid evidence of Michelangelo intentionally concealing anatomical structures in the Sistine ceiling is scant. Sceptics such as Salcman suggest that proponents of the hidden anatomy theory may simply be seeing physical structures where there are none, superimposing meaning on ambiguous patterns in a way similar to the phenomenon of pareidolia, where humans are primed to perceive faces in inanimate arrays of features. Indeed, some have claimed that God's neck shows a goitre, rather than a brainstem, while others perceive a bisected right kidney in the *Separation of Land and Waters*.

The idea of hidden depictions of anatomical structures in the works of Renaissance masters is an intriguing one, but in the absence of stronger evidence or a more compelling rationale for these inclusions, it may be the case that the only thing revealed in these cases is a penchant for anatomical pareidolia on the part of the viewer.

Notes

1 Described by Galileo in 1609 in a letter to the Doge of Venice thus:

> 'The power of my cannocchiale [telescope] to show distant objects as clearly as if they were near should give us an inestimable advantage in any military action on land or sea. At sea, we shall be able to spot their flags two hours before they can see us; and when we have established the number and type of the enemy craft, we shall be able to decide whether to pursue and engage him in battle, or take flight. Similarly, on land it should be possible from elevated positions to observe the enemy camps and their fortifications.'

2 It is rumoured, however, that as he recanted he was heard to mutter 'E pur si muove' ('It does move, though') in reference to the Earth moving around the sun and not vice versa.

3 We are reminded of the words of Theodore Roszak (1972): 'nature composes some of her loveliest poems for the microscope and the telescope'.

4 The Neuron Doctrine was unequivocally confirmed in the 1950s following the invention of electron microscopy, which allowed for the first visualisations of connectivity between brain cells.

5 Aptly, the European Space Agency probe launched in 1985 to study the comet was named Giotto in his honour.

6 For more on Halley's Comet, see Episode 4 ('Heaven and Hell') of Carl Sagan's *Cosmos* series.

7 See Olson, Doescher and Olson (2004) and Zerefos, Gerogiannis, Balis, Zerefos and Kazantzidis (2007) for a more detailed discussion.
8 Ashrafian (2013).
9 For further description and discussion, see Lazzeri and Nicoli (2016).
10 See Lazzeri, Castello, Matucci-Cerinic, Lippi and Weisz (2016).
11 See Suk and Tamargo (2010) for a discussion.

References

Ashrafian, H. 2013. Leonardo da Vinci and the First Portrayal of Pectus Excavatum. *Thorax*. 68(11), p. 1081.

Cajal, S.R. 1996. *Recollections of My Life*. Cambridge, MA: MIT Press.

Galilei, G. 1967. *Dialogue Concerning the Two Chief World Systems, Ptolemaic and Copernican*, trans. by Stillman Drake. Berkeley, CA: University of California Press.

Galilei, G. 2016. *Sidereus Nuncius, or The Sidereal Messenger*, trans. Albert Van Helden. Chicago, IL: University of Chicago Press.

Lazzeri, D. and Nicoli, F. 2016. Pectus Excavatum in Paintings by Jusepe de Ribera (1591–1652). *Thorax*. 71(7), pp. 669–670.

Lazzeri, D., Castello, M. F., Matucci-Cerinic, M., Lippi, D., and Weisz, G. M. 2016. Osteoarthritis in the Hands of Michelangelo Buonarroti. *Journal of the Royal Society of Medicine*. 109(5), pp. 180–183.

Olson, D.W., Doescher, R.L., and Olson, M.S. 2004. When the Sky Ran Red: The Story Behind The Scream. *Sky Telescope*. 107(2), pp. 29–34.

Suk, I., and Tamargo, R. J. 2010. Concealed Neuroanatomy in Michelangelo's Separation of Light From Darkness in the Sistine Chapel. *Neurosurgery*. 66(5), pp. 851–861.

Zerefos, C. S., Gerogiannis, V. T., Balis, D., Zerefos, C. S., and Kazantzidis, A. 2007. Atmospheric Effects of Volcanic Eruptions As Seen By Famous Artists and Depicted in Their Paintings. *Atmospheric Chemistry and Physics*. 7, pp. 5145–5172.

4

A THOUSAND DATA POINTS

Art in scientific visualisation

Introduction

Previous chapters have illustrated how scientists – da Vinci, Vesalius, Galileo, Cajal – used their artistic ability to convey scientific observations or principles using their drawings, etchings and paintings to show others what they had seen. In many ways, these representations mark the birth of science communication, the process whereby scientists attempt to explain their findings to society at large. Yet, seminal as they were in spreading scientific ideas to a wider audience, these renderings typically depicted *only one case* – a single observation, a single view through the telescope, a single dissected organ. As the relationship between art and science matured, so too did the content of such representations evolve: they began to depict multiple observations, entire datasets – even phenomena and ideas which could not actually be seen. Further, the means used to create such images grew beyond drawings and paintings, becoming more abstract, more flexible but – as we shall see – no less beautiful.

We are now more adept at representing complex information in visual form than at any stage in the history of our species; advances in technology and computing have allowed us to depict enormous collections of observations in increasingly elegant and inventive ways. The heavy reliance of our species on vision (over 55% of the primate brain is devoted to this sense) has resulted in the evolution of a system which rapidly perceives, processes and interprets visual information, allowing us to understand almost instantly a single snapshot depicting literally billions of data points and the patterns they convey. In this way, the glowing lines in Figure 4.1 can be used to represent cycle patterns in Vienna; the brightest lines showing the routes that are used most often. Similarly, a morphed, distorted map of the globe can depict the populations of different countries (see Figure 4.2). We have become so expert at generating creative ways to convey large-scale observations that their aesthetic and innovative merits are celebrated in collections such as *Information is Beautiful* and *Knowledge is Beautiful*. As a result, visualised information has become a new artform[1].

Information as art

Although we are accustomed to seeing and interpreting images such as those in Figures 4.1 and 4.2 today, the ability to represent large datasets or collections of observations in a single image is a relatively recent development in the history of science. We have already seen the powerful influence the anatomical and astronomical illustrations of da Vinci, Galileo, Cajal

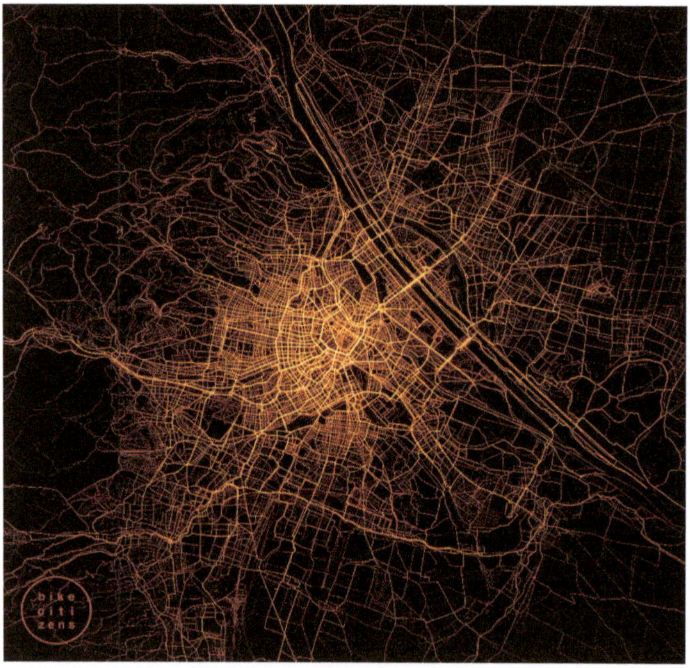

FIGURE 4.1 Heatmap of Vienna created through cycling with the Bike Citizens App (Copyright Bike Citzens).

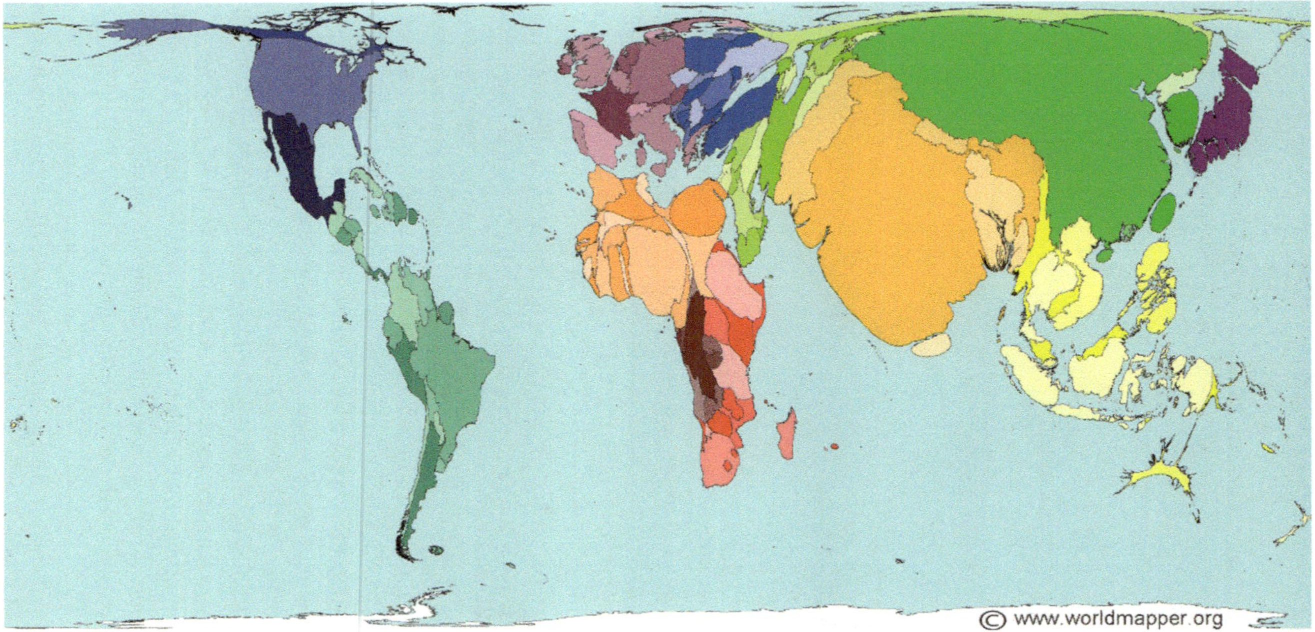

FIGURE 4.2 The size of each territory shows the relative proportion of the world's population living there (Copyright ©Worldmapper.org/Sasi Group [University of Sheffield] and Mark Newman [University of Michigan]).

FIGURE 4.3 Illuminated copper engraving from *Metamorphosis insectorum Surinamensium*, Plate XLV. 1705 by Maria Sibylla Merian (1647–1717) [CC BY 3.0], via Wikimedia Commons.

and others had on the scientific knowledge of their respective eras, and the importance of artistic depictions of single cases were a common occurrence in the development of many fields since the Renaissance. The seventeenth-century German illustrator and naturalist Maria Merian (1647–1717) made an enormous contribution to the field of entomology, bringing her artistic talents to bear in the service of science by producing drawings which revealed the natural process of butterfly metamorphosis – at this time, insects were considered evil, and the change from caterpillar to winged butterfly was viewed as a supernatural act (see Figure 4.3). Even more remarkable is the fact that these illustrations were produced when Merian was only 13 years of age. In Ireland, Ellen Hutchins (1785–1815) made equally impressive contributions to the field of botany. In her short life, Hutchins produced over 400 drawings of local plant specimens (Figure 4.4), discovering several new species of moss and seaweed in the process. In the United States, the field of ornithology was greatly advanced by the sumptuous illustrations of 130 species of birds' nests and eggs by Genevieve Jones (Figure 4.5; 1847–1879) and, after Genevieve's untimely death from typhoid at the age of 32, by her mother Virginia (published in *Illustrations of the Nests and Eggs of Birds of Ohio*) (Popova, 2012)[2]. Similarly, the little-known paintings and drawings of fungi and mushrooms by the artist and author Beatrix Potter (Figure 4.6; 1866–1943) would eventually lead to advances in the understanding of the classification of lichens as a fungus–algae hybrid[3]. Had her observations not been dismissed due to her sex, her impact on the ensuing debate regarding

FIGURE 4.4 Seaweed specimen collected by Ellen Hutchins. Image courtesy of the Herbarium, Trinity College Dublin.

FIGURE 4.5 Plate II. *The Nest of the Wood Thrush* by Genevieve Jones, Courtesy of the Smithsonian Institution Libraries, Joseph F Cullman 3rd Library of Natural History. [Public domain or CC0], via Wikimedia Commons.

FIGURE 4.6 *Hygrophorus Puniceus* by Beatrix Potter, 1894, watercolour. Collected at Smailholm Tower, Kelso. Courtesy of the Armitt Trust, the Armitt Museum and Library, Ambleside.

the status of lichen would have paralleled Cajal's decisive role in the acceptance of the Neuron Doctrine.

While these contributions would continue to inform and enrich the sciences to which they contributed, parallel strands of data visualisation were beginning to emerge, strands whose roots extend back to the earliest records of our species. These were attempts to convey information arising from multiple observations or to visualise ideas or concepts that could *not* be seen. Both endeavours would require a step towards abstraction and a leap of imagination by those who pioneered them.

Mapping the path of visualisation

Maps are among the oldest and most elegant means devised to convey a wealth of information in a single picture. To draw a map of the environment, to represent in a two-dimensional image the spatial relations that exist between features in a three-dimensional world, demands complex cognitive operations and no small amount of skill. The ability to extract ourselves from the first-person perspective through which we normally view our surroundings, to mentally elevate ourselves above and beyond our typical position to identify the broader patterns and distribution of the geographical landmarks that surround us, requires a level of sophistication seen in few species. Tracing the history of map-making tells us much about not only the origins of data visualisation, but also about the cognitive development of humans from our earliest beginnings.

It has been suggested that primitive cave art may reveal our ancestors' earliest attempts to depict concrete objects – animals, people – and their salient features in visual form, and we will see later in Chapter 6 that such cave paintings may also hint at something of the psychological experience of these early humans. But the images found in our ancient homes also show evidence of our first attempts at generating maps; patterns of dots found on the walls of caves at Lascaux and Cuevas de El Castillo are probably 8,000-year-old astrological maps charting the relative positions of specific stars or constellations. Also present in some of these caves are crude representations of dwellings and villages, though this has been disputed by some. The maps drawn by the earliest civilisations – Ancient Babylonians, Egyptians – often focused on small regional locations, such as a river valley (24th/25th century BC), the holy city of Nippur (14th/15th century BC) or the area east of the Nile (the Turin Papyrus Map, c. 1160 BC), with simple features like hills, rivers and cities outlined.

Early world maps – mainly from Ancient Greek map-makers, including Anaximander (611–546 BC) and Hecataeus (550–475 BC) – often conceived of Earth as being circular or cylindrical, surrounded by a peripheral ocean. Building on the work of Pythagoras and Eratosthenes, it was Claudius Ptolemy (c. AD 100–c. 170) who produced the first depiction of the world as spherical, with accurate locations of geographic features. Ancient Rome also had its map-makers, again producing both global (e.g. the world map of Pomponius Mela, 43 AD, which divided Earth into five regions) and local (e.g. the Tabula Peutingerania, a fifth-century map of the Roman roads) representations. Similar endeavours were being carried out in China (Pei Xiu is considered the 'Ptolemy of China'), Mongolia, India and as far afield as Polynesia. This persisted into the Middle Ages, when the Mappa Mundi was produced, a symbolic depiction of the Earth as circular, symmetrical and again surrounded by an encircling ocean.

The detail and sophistication of maps underwent significant advancement from the Renaissance onwards, driven largely by necessity – the drive for trade and exploration gave rise to a need for accurate navigation charts and maps of sea routes. This resulted in the emergence of the Italian[4], Iberian, Dutch and German Cartographic Schools, producing maps and charts with

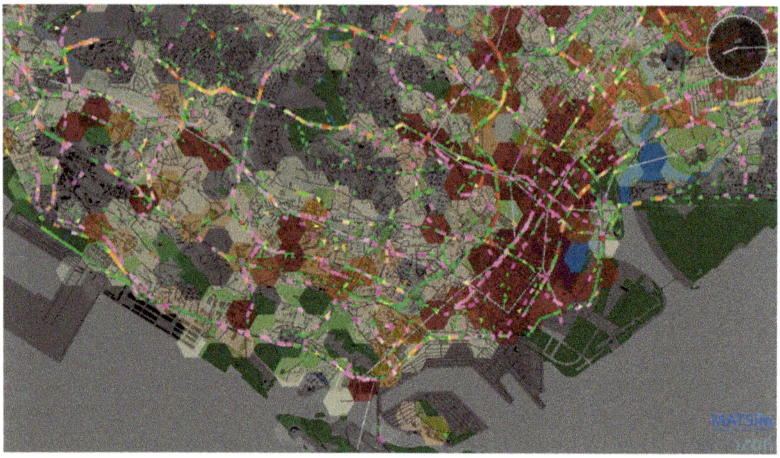

FIGURE 4.7 The agent-based transport simulation, MATSim Singapore, models the activity and travel decisions of the entire commuter population of the city state. Credit: Future Cities Laboratory Engaging Mobility project and senozon AG.

greater detail and accuracy than had ever been seen before as the new discoveries and knowledge gleaned from each voyage were fed back into the next generation of maps. This continued into the eighteenth century and the 'Golden Age' of modern cartography (1872–1945), during which novel ways were devised to present what was known about the geography of the world, the distended-looking Homolosine (Goode, 1923) and Loximuthal (Sieman, 1935) Projections among the prominent twentieth-century advancements.

The advent of new technologies in the 1970s and 1980s allowed for further advancements; most notably, the emergence of 'Space Syntax' – a scientific field dedicated to investigating the relationship between urban spatial layouts and how these influence societal behaviour and environmental phenomena[5]. The interplay between space and human activity can clearly be seen in the stunning maps produced by this approach (see Figure 4.7 for an example).

While maps of the physical environment hold an enduring appeal for both their practical and aesthetic merits, pictorial representations of imaginary landscapes and fictional locations also have the ability to fascinate and enthral us[6]. From maps of the lost city of Atlantis to Dante's levels of hell to Tolkien's map of Middle Earth, a map – whether real or fictional – represents a way to convey a vast amount of geographical (and in some cases sociological or political) information in a single, beautiful image. Just as Greece was invariably at the centre of early depictions of the Earth, maps occupy a special, pivotal place in the pantheon of data visualisation.

From nature's web to the Tree of Life: von Humboldt and Darwin

When we consider the history of representing large quantities of information graphically, two nineteenth-century naturalists stand head and shoulders above most others; Alexander von Humboldt and Charles Darwin. Both found novel, innovative and even visually pleasing ways to describe their voluminous observations simply and accessibly, drawing on the very natural world that they studied as a means for conveying their ideas.

Alexander von Humboldt (1769–1859) was born in Berlin into a family of Prussian aristocracy, but his innate curiosity would drive him far afield to become one of the great explorers and naturalists of his day, and one who would, in turn, inspire Charles Darwin to embark on his voyage on the Beagle. An adventurer and explorer, von Humboldt was a forward thinker on topics including science communication, the links between art and science, and the interconnectedness of natural systems, proposing many views which would prove decades, even centuries, ahead of their time[7]. He famously conceived of nature as being akin to a web, where every species and system – all organisms, all aspects of environment and climate – were co-dependent, co-related, such that any perturbation of one would be felt among the others. As he explained himself:

> In this great chain of causes and effects, no single fact can be considered in isolation.
>
> *von Humboldt and Bonpland (2009, p. 79)*

A beautiful example of his ability to illustrate his observations came during a five-year expedition to South America. In 1800, shortly after his arrival, von Humboldt witnessed first-hand how the clearing of forests and diversion of water had led to the lowering of the water levels of Lake Valencia, turning the once lush green land into a dry, barren place. Taking extensive measurements, he concluded that deforestation had caused the nutrients to be washed away from the soil, leaving it infertile. This, in turn, prompted farmers to cut down more of the forest to grow their crops, and so the cycle of destruction continued. Inspired by his findings and eager to continue this line of research, von Humboldt travelled further into the Andes to the famous mountain Chimborazo.

Beginning at the foothills of this great peak, he sketched a cross-section of the entire mountain, but this was to be no ordinary painting; rather, it was a scientific illustration. He showed the distribution of plants, lichens and fungi and mapped them onto his painting according to where he found them while also indicating their corresponding altitudes. He recognised that plants and species had a particular home on different parts of the mountain. He observed Alpine plants, oaks and ferns, which were recognisable to him from their presence in Europe, all located according to altitude and temperature. It dawned on him that there is a unity in the world of nature: that climate, plants and species were all connected, even across continents – all separate strands of his web of nature, yet all tied together. The simplicity of this idea is conveyed by the richness of data presented in his elegant illustration (Figure 4.8).

The scientific world of the 1800s was dominated by the categorisation of plants and species according to their taxonomy, emphasising the differences between them and intent on marking the boundaries which separated them. It took von Humboldt to see the larger picture from a different angle, focusing instead on the interconnected relationships flora and fauna have with other aspects of nature. His painting of Mount Chimborazo he called *Naturgemalde,* which roughly translates from German to 'painting of nature', and his use of such a means to convey his findings reflects his distinct view on the link between art and science. Further, as mentioned, his metaphor of a web to describe the connectedness of nature would greatly influence Charles Darwin, who later chose another natural form, a tree, to encapsulate his views on the origins of species.

Having completed his studies of botany and geology at the University of Edinburgh and then Cambridge, the great naturalist Charles Darwin (1809–1882), inspired by the works and expeditions of von Humboldt, embarked on a five-year (and one day) voyage aboard the HMS Beagle which would change his life and our understanding of the relationships between species. He was offered a place on the ship thanks to a recommendation from his former mentor John Henslow, and the expedition to survey South America departed in December 1831. The route took him along the east coast of the continent, rounding the southernmost tip at Patagonia in 1833, navigating up the west coast the following year and included – crucially – a five-week stay in the Galápagos in 1835. Here, his exposure to the flora and fauna of these recently created volcanic islands – in particular the variety of finches, giant tortoises and iguanas – would later inspire and inform his understanding of the relations between different and similar species.

Two years later, when he was in the process of formulating his theory of evolution, he – just as von Humboldt had before him – would employ a natural form, a tree, as a metaphor to explain the way in which species arising from a common ancestor either adapt and proliferate, or fail to evolve and perish. His initial rudimentary sketch of this 'Tree of Life', accompanied by

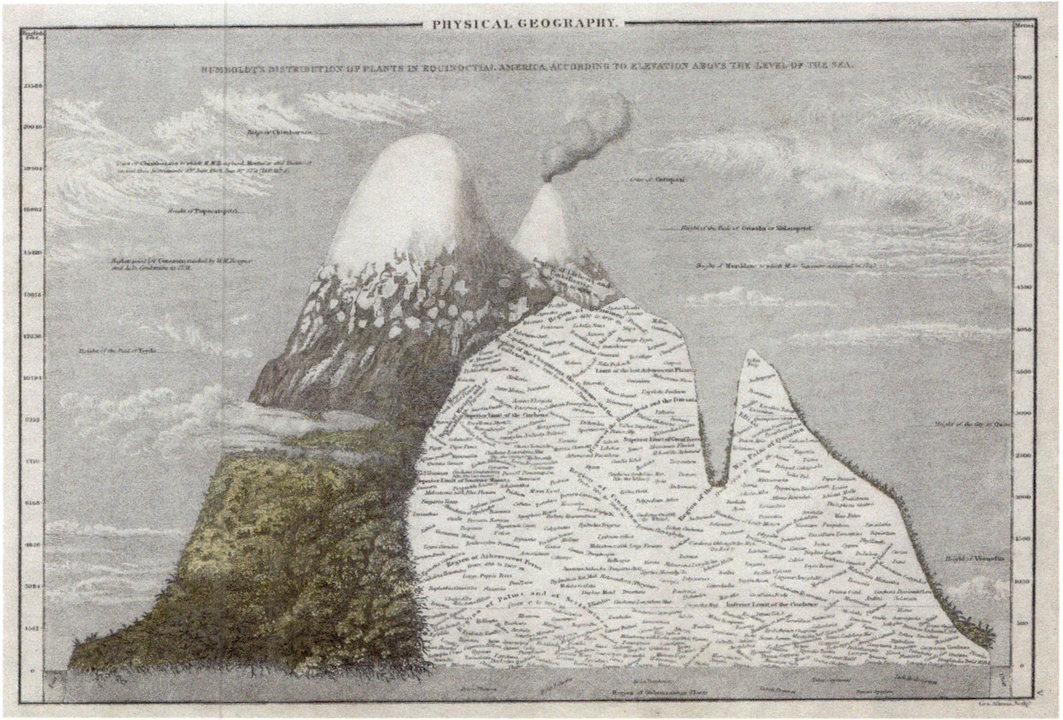

FIGURE 4.8 von Humboldt's illustration of Chimborazo mountain in the Andes. (By Bonpland, Aimé, Arzt, Naturforscher, Entdeckungsreisender, Frankreich, 1773–1858 [Zentralbibliothek Zürich] [Public domain], via Wikimedia Commons.)

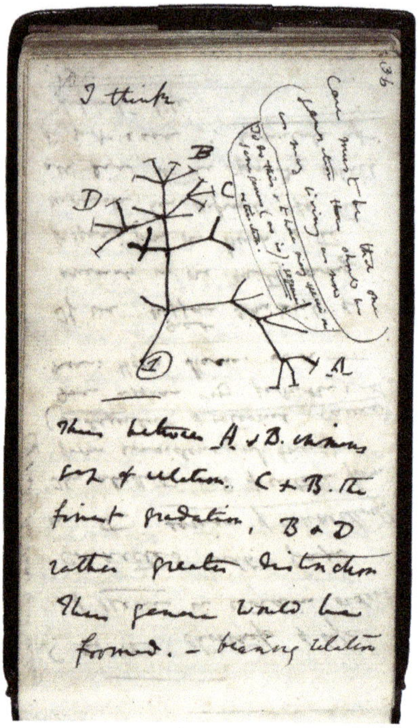

FIGURE 4.9 Darwin's original Tree of Life illustration [Public domain], via Wikimedia Commons.

the simple preface 'I think' (Figure 4.9) would later be replaced by a more sophisticated (and considerably neater) version when he finally published *On the Origin of Species* in 1859. Although he was not the first to use a tree as a means of explaining how different organisms may be related to one another[8], Darwin's 'Tree of Life' is undoubtedly the best known. The image simultaneously depicts the relationships of different groups of surviving animals along the densely sprouting branches, while also explaining the fate of extinct creatures – via his proposed mechanism of natural selection – using curtailed or pruned limbs. And although his first sketch of the diagram was crudely rendered and absent of most aesthetic merit, when viewed in combination with Darwin's written explanation, the elegance and power of the Tree of Life as a means of articulating his many observations truly emerges:

> As buds give rise by growth to fresh buds, and these, if vigorous, branch out and overtop on all sides many a feebler branch, so by generation

> I believe it has been with the great Tree of Life, which fills with its dead and broken branches the crust of the earth, and covers the surface with its ever branching and beautiful ramifications.
>
> *Darwin (1859, p. 130)*

The Nightingale and the roses

Around the same time that Darwin was conceiving his Tree of Life, one of the first attempts to illustrate a pattern of observations graphically in the field of medicine was made by his compatriot, Florence Nightingale (1820–1910). Although she is best known throughout the world for her work as a nurse and a caregiver, her contribution to statistics and data representation easily matches her humanitarian efforts.

At the age of 24, flouting the convention of the time, she resolved to become a nurse and proceeded to educate herself to this end. With the outbreak of war in Europe in 1853, Nightingale was sent with a team of nurses to a British base near Istanbul. What awaited her was an appalling situation; inadequate care for the wounded, poor hygiene and limited medical staff and supplies all contributed to a staggering death toll. While working tirelessly to tend to the injured and dying, she took note of both the number of deaths that occurred and their causes. During the first winter of the war (1853/1854) more than 4,000 soldiers died, but Nightingale observed that between *6 and 10 times* the number of deaths resulted from infections and/or poor hygiene than were directly due to wounds sustained in battle. She wrote to *The Times* in London reporting these observations and lobbied parliamentarians for improved conditions in the wards. Six months after her arrival in Turkey, the British Government despatched the Sanitary Commission to the field hospital, with a mandate to set about flushing out the sewers and improving ventilation in the wards; an immediate decline in the mortality rate followed.

After the war ended, a Commission was set up in London to investigate sanitary conditions in the army, and this Commission was to be informed by the statistical records and notes Nightingale had kept during her time in Turkey. Rather than supplying a large, tabulated dataset with long lists of numbers and figures – tables which the politicians might misunderstand or misinterpret – she chose instead to illustrate her data graphically in what we now refer to as the 'rose diagram'. She produced a polar area diagram which showed the number of deaths that occurred due to disease, wounds and other causes; it did so simply, dramatically and beautifully. Figure 4.10 shows that more soldiers died of disease (blue area) than from the wounds they had received during battle (red area). In addition, comparison with the second

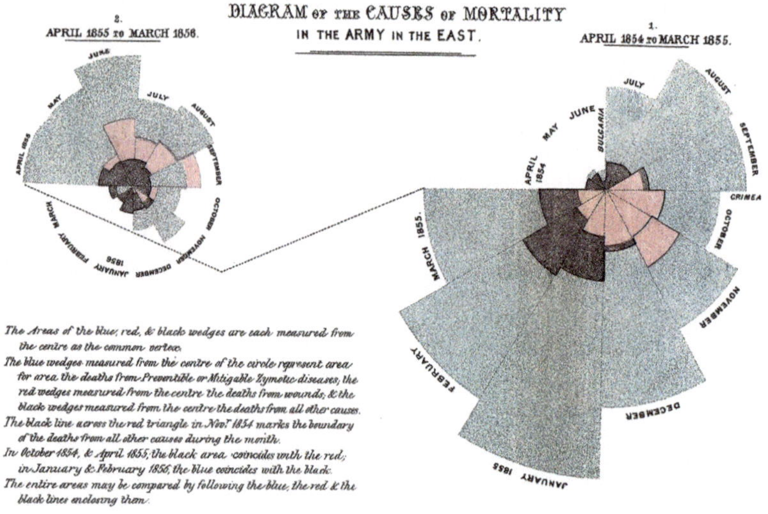

FIGURE 4.10 Nightingale's Rose Diagrams on the causes of mortality among soldiers, 'Diagram of the causes of mortality in the army in the East', published in *Notes on Matters Affecting the Health, Efficiency, and Hospital Administration of the British Army*. (By Florence Nightingale [1820–1910]. [Public domain], via Wikimedia Commons.)

polar – or 'rose' – illustrates a dramatic decrease in disease-related deaths, an improvement attributed to better health care and improved sanitary conditions after the intervention of the Sanitary Commission. With these rose diagrams, Nightingale took a large dataset and transformed it into a simple, intuitive figure conveying the patterns of her observations. Moreover, she opened the door to new means of visualising scientific data in creative and effective ways.

Turing's enigmatic drawings

While von Humboldt, Darwin and Nightingale grappled with and overcame the challenges of embodying many observations in a single image, others would confront a different but equally vexing problem – how to convey ideas or phenomena which the eye could not (now or perhaps *ever*) see. In the later years of his career, Alan Turing (1912–1954), the revolutionary computer scientist and World War II code-breaker, faced such a problem when he investigated the mathematical basis underlying morphogenesis, the mechanism by which biological organisms take certain shapes. He had observed

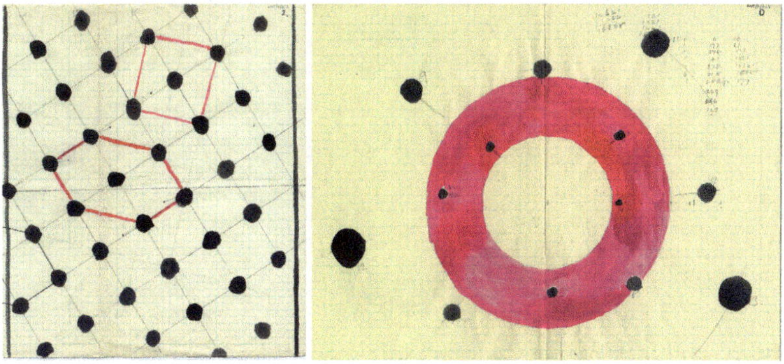

FIGURE 4.11 Coloured diagrams showing patterns of dappling and calculations, made by Alan Turing in his work on morphogenesis: (left) a parastichy diagram, (right) the daisy ring diagram (n.d. Paper, 8 sh. in envelope; Copyright © PN Furbank).

the presence of certain mathematical algorithms and patterns – such as the Fibonacci sequence – in biological structures like daisies and sunflowers, and correctly surmised that the diffusion of chemical signals could be predicted and modelled mathematically, and that fluctuations in these patterns would determine the development of particular morphologies in different organisms. To help him conceptualise these interactions, Turing produced many coloured illustrations; blobs, lines, connected circles, coloured rings – all hand-drawn on sheets of graph paper – representing dappling patterns, leaf arrangements and daisy rings (Figure 4.11). In this manner, his grasp and use of abstraction facilitated his understanding of one of the most concrete of phenomena, the very *shape* of things in the natural world. Turing's pioneering work on morphology would be borne out decades later, confirmed by the elucidation of the structure of DNA by Watson, Crick, Franklin and Wilkins in 1953. In fact – not unlike Turing's drawings – it was Franklin and her PhD student Raymond Gosling's X-ray diffraction photographs of DNA taken one year previously which sparked the very idea that DNA could be helical (Figure 4.12).

Ode to a diagram: Richard Feynman's visual notation

In a similar way, the Nobel Prize-winning scientist, bongo player and amateur artist Richard Feynman (1918–1988) utilised a means of visualisation to better explain the nature of how subatomic particles interact. Like da Vinci

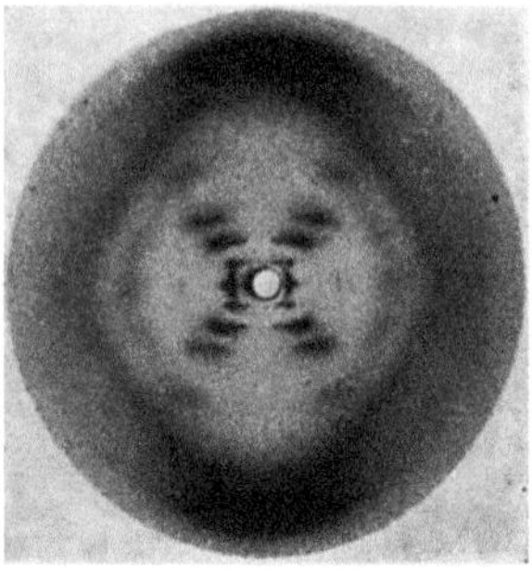

FIGURE 4.12 Franklin and Gosling's diffraction photograph, 'Photo 51' of DNA structure (by source [WP:NFCC#4], fair use).

and von Humboldt, Feynman understood the value of aesthetic considerations and how artistic endeavour can be used in the service of science:

> I wanted very much to learn to draw, for a reason that I kept to myself: I wanted to convey an emotion I have about the beauty of the world … there's a generality aspect that you feel when you think about how things that appear so different and behave so differently are all run 'behind the scenes' by the same organization, the same physical laws. It's an appreciation of the mathematical beauty of nature, of how she works inside; a realization that the phenomena we see result from the complexity of the inner workings between atoms; a feeling of how dramatic and wonderful it is. It's – of scientific awe – which I felt could be communicated through a drawing.
>
> *Feynman (1985, p. 261)*

Following on from the work of predecessors such as Ernst Stueckelberg, in 1948 Feynman invented a new notation to represent how minuscule particles like electrons and photons interact as they move towards each other on collision courses and subsequently scatter. His clear, intelligible diagrams (Figure 4.13) – to which his name has become attached – not only allow the nature of these quantum interactions to be better visualised or imagined, they allow the specific probabilities of different routes to different outcomes to be calculated. So pivotal has been the impact of these diagrams that, almost 70 years after their introduction, much of the work currently taking place at the Large Hadron Collider (LHC) at Cern relies heavily on them[9]. In the

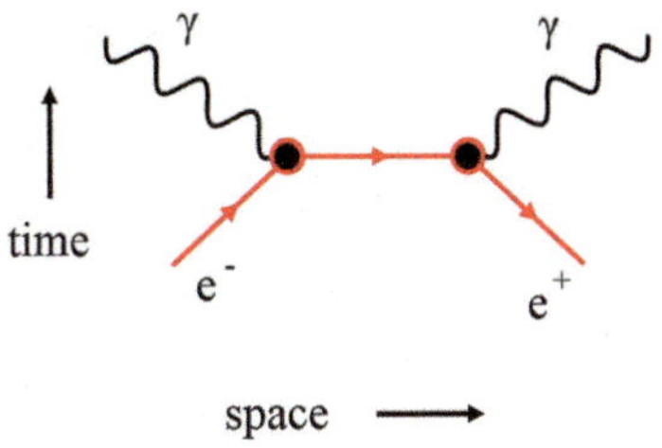

FIGURE 4.13 Feynman diagram showing Space–Time vectors of Electron–Positron Annihilation (Public domain).

same way that Turing represented his morphogenetic hypotheses, Feynman's drawings afforded a way to picture and conceptualise something not merely unseen, but *unseeable*.

Somewhere over the brainbow

The examples described so far in this chapter have demonstrated how large datasets containing many thousands of observations can be represented in simple, clever and accessible ways using visualisation techniques. But when we consider large numbers of data points, no structure presents a greater challenge than the human brain: depicting an organ with close to a hundred billion cells making trillions of connections would require the invention of novel approaches to cell visualisation which go far beyond the wildest imaginings of Cajal or Golgi. While many brain-mapping methodologies used in neuroscience (like electroencephalography [EEG] and neuroimaging) focus on electrical or metabolic correlates of brain function, structural approaches are concerned with visualising different cell types, their structure and connectivity. Few such approaches generate images as arresting and aesthetically pleasing as the *brainbow* (Figure 4.14).

Developed by Roger Tsien, and building on pioneering work by Osamu Shimomura and Martin Chalfie (the trio shared a Nobel Prize in 2008 for this work), the brainbow technique makes use of glowing fluorescent proteins (GFPs) and the gene that gives rise to them. The first two such proteins – green and subsequently blue – were discovered by Shimomura in the 1950s and 1960s in his attempts to explain the bioluminescence of specific species of jellyfish. Upon learning about these proteins at a scientific conference in the late 1980s, Chalfie introduced the gene responsible for producing these GFPs into the nervous system of the simple transparent

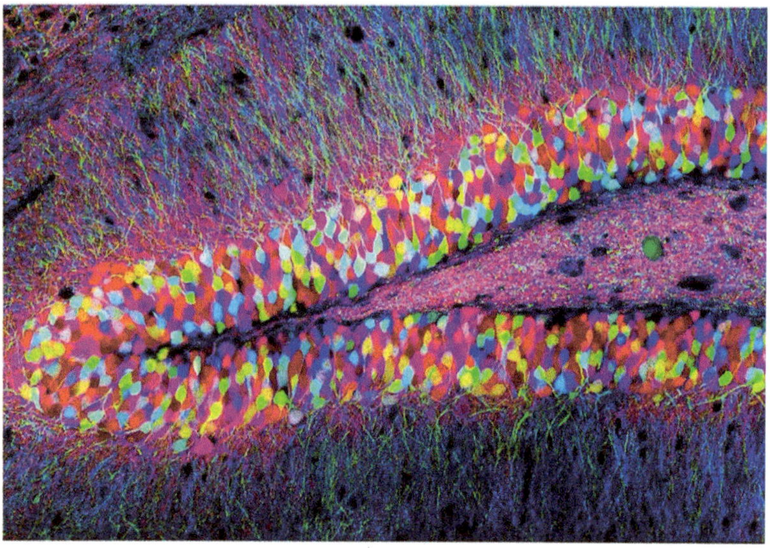

FIGURE 4.14 The mouse hippocampus using the brainbow technique. Image by Tamily Weissman. The Brainbow mouse was produced by J Livet, TA Weissman, H Kang, RW Draft, J Lu, RA Bennis, JR Sanes and JW Lichtman. *Nature* (2007) 450:56–62.

type of worm (*C. elegans*) which he was studying at the time. This allowed him to observe for the first time the different cells of the worm's nervous system glowing due to the presence of the protein; further, he could track the changes in the machinery of the fluorescent-tagged cells as the worm went about its business.

Taking the technique a step further, Tsien produced other fluorescent proteins in different colours, giving him a palette with which to visualise the myriad cell types of a far more complex organ in a much more sophisticated animal – the human brain. This is a striking example of how fundamental discoveries (e.g. in the jellyfish) can have far-reaching, broader consequences for many species, including ourselves, thereby underscoring the importance of basic research in science. His brainbows allowed multiple cells, thousands of neurons and whole fibre tracts to be visualised and tracked at the same time in a rainbow of colours. With these remarkable discoveries, scientists can now track multiple cancerous tumours, identify different cells that may be infected by disease and gain a greater understanding of cellular processes. The beauty of these images is matched by their scientific value.

Mapping brain activity using these and other sophisticated imaging techniques is one of the fastest-growing research fields in science (known as 'Connectomics'). This is the primary aim of the Human Connectome Project. The foundation of this initiative is the 'connectome', a complex matrix of the structural connections within the nervous system, defined at varying levels of analyses from macro-connections (between brain regions) to nano-connections (between nerve cells)[10]. An important by-product of mapping 'normal' brain activity is that it can be compared across different patient groups in an attempt to understand how brain structures might differ in their response to stimuli.

In the project, thousands of brains are scanned to trace 'normal' connectivity and then this pattern is compared with those of patients with specific illnesses to see if key areas within the brain are wired differently (named 'connectopathies'). The thinking behind this project is that brain disorders are disorders of connectivity between important structures rather than dysfunction within a single structure.

The project relies on a technique called Diffusion Spectrum Imaging (DSI), whereby the direction of movement of water molecules in the brain can be imaged. As most water molecules move along nerve fibres (like drainpipes), thousands upon thousands of nerve tracts can be imaged and visualised. What emerges is a stunning and colourful illustration of the architecture of functional connectivity within the brain (Figure 4.15), the intricate

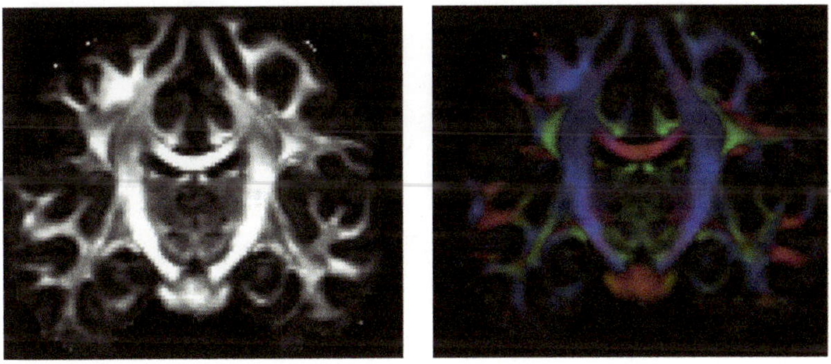

FIGURE 4.15 Fractional anisotropy (left), and principal diffusion directions (right) images from the HCP dMRI data provide a measure of how water diffuses in the brain. Diffusion directions are RGB-colour encoded – red: left–right, green: anterior–posterior, blue: inferior–superior. (Image courtesy of the WU-Minn HCP consortium.)

and staggering proliferation of billions of cells collected into elaborate highways of communication; a technicolour map of a living brain in action.

As the purpose of visually depicting scientific ideas shifted from representing single cases to conveying multiple observations and datasets in a sophisticated yet efficient way, the need for artistic ability was, in many ways, replaced by an appreciation for visual aesthetics. In creating their striking diagrams, Nightingale, Darwin and the others described in this chapter were forced to confront a problem common to those who engage in science communication – the trade-off between presenting a simple and understandable visual representation and the need to include sufficient detail to ensure the viewer is not misled. In the same way that presenting an analogy for a complex idea runs the risk of oversimplification, so too must data visualisation balance the twin needs for elegance and accuracy. The examples described here skilfully negotiate this challenge, combining scientific integrity with aesthetic and intuitive appeal.

One of the key differences between visual art and science centres around the issues of interpretation and subjectivity – a work of art is inherently personal, subjective and open to different interpretations on the part of the viewer. In science, the aim is clarity, objectivity and the absence of subjective interpretations. It is somewhat ironic, then, that some of the visual depictions generated in the service of science are met with a similar reaction of aesthetic appeal as some artistic creations, despite the two being at cross purposes.

Notes

1 See Finn (2012).
2 For more information on their sad story, see Maria Popova's 'A Radical Journey of Art, Science, and Entrepreneurship: A Self-Taught Victorian Woman's Visionary Ornithological Illustrations'.
3 See Lear (2016).
4 Including one by da Vinci – see Tyler (2017).
5 For more information on Space Syntax, see Hillier and Hanson (2008).
6 See Eco (2008) for more on this topic.
7 See Wulf (2015) for more details on the life of Alexander von Humboldt.
8 In fact, the use of this metaphor is very old; see Pietsch (2012) for a detailed overview of the history of its usage.
9 See 'Feynman Diagrams: Taking a closer look at LHC' on the Large Hadron Collider (LHC) website.
10 For an excellent review of the connectome approach and its future directions, see Swanson and Lichtman (2016).

References

Darwin, C. 1859. *The Origin of Species Or, The Preservation of Favoured Races in the Struggle for Life*. London: John Murray.

Eco, U. 2008. *The Book of Legendary Lands*. London: MacLehose Press.

Feynman, R.P. 1985. *"Surely You're Joking, Mr. Feynman!": Adventures of a Curious Character*. New York: WW Norton & Company.

Finn, H. 2012. Making Data Beautiful. *The Wall Street Journal* (6 April).

Hillier, B., and, Hanson, J. 2008. *The Social Logic of Space*. Cambridge: Cambridge University Press.

Lear, L. 2016. *Beatrix Potter: A Life in Nature*. New York: St. Martin's Griffin.

Pietsch, T.W. 2012. *Trees of Life: A Visual History of Evolution*. Baltimore, MD: The Johns Hopkins Press.

Popova, M.A, *Radical Journey of Art, Science, and Entrepreneurship: A Self-Taught Victorian Woman's Visionary Ornithological Illustrations*, www.brainpickings.org/2012/06/27/americas-other-audubon.

Swanson, L., and Lichtman, J. 2016. From Cajal to Connectome and Beyond. *Annual Review of Neuroscience*. 39, pp. 197–216.

Tyler, C. 2017. Leonardo da Vinci's World Map. *Cosmos and History: The Journal of Natural and Social Philosophy*. 13(2), pp. 261–280.

von Humboldt, A. and Bonpland, A. 2009. *Essay on the Geography of Plants*, ed. and trans. by Jackson, S.T. Chicago, IL: The University of Chicago Press.

Wulf, A. 2015. *The Invention of Nature*. London: John Murray.

5
THE PEOPLE BEHIND THE DATA

Introduction

The examples described in preceding chapters illustrate how visual representations in science evolved from depicting single cases to multiple observations, ranging from individual specimens (in natural and anatomical drawings) to hundreds (in the case of rose diagrams), thousands or even billions (as in brain imaging) of data points. And while we may rightly marvel at the skill and ingenuity involved in these depictions of various states of nature, it can be all too easy to lose sight of an important fact about such images: that, very often, the data points illustrated represent individuals or groups of individuals. Sometimes they convey entire species of flora or fauna and their fate (as in the Tree of Life), but often they show *people* – human beings or some aspect of their lives. In the case of Nightingale's rose diagrams, it was their health status. In medical case studies, such as the so-called Elephant Man Joseph Merrick, a unique or rare malady is described. In this chapter, we recount the lives of two famous case studies from the history of neuropsychology. Although separated by a century, these two men share the unusual distinction of having been – until very recently – more easily identifiable by their brains than their faces. Their stories act as a powerful reminder that, while information can be and often *is* beautiful, we must strive to remember that there are people behind the data points.

The history of neuroscience is punctuated by reports and case studies of unfortunate individuals who – through accident or, occasionally, design – sustained damage to a region of their brain. Survivors of such events have traditionally been of enormous interest to the medical community, not only for the fascinating or sometimes bizarre deficits resulting from their tissue loss, but equally for the capacities which remain intact following the injury.

The earliest written reference to the 'brain' in the human record can be found in the Edwin Smyth Surgical Papyrus, an ancient Egyptian medical text detailing 46 case studies of physical injury, 23 of which involve the brain. The stroke patients of Pierre-Paul Broca (1861) and Karl Wernicke (1874) afforded us an unprecedented insight into the workings of the language system. Gunshot wound victims following each of the World Wars provided new insights into how visual, auditory, tactile and attentional functions operate. But of all the thousands of neuropsychological case studies that have been reported, two individuals – Phineas Gage and Patient HM – stand out from the others; this is due in part to the strangeness of their stories, but also for the enormous contribution to knowledge with which their remarkable conditions provided us.

A portrait of the mind

Phineas P Gage (1823–1860) was a railroad worker whose job it was to oversee the laying of a new track for the Rutland and Burlington Railroad through the state of Vermont. It was a major undertaking, and one aspect of Gage's position involved the blasting of rocks from the proposed route of the tracks. A hole would be drilled into the offending outcrop, explosive powder poured into the cavity, then sand poured on top and the whole cocktail pushed deep into the rock using a long metal rod called a tamping iron. The rod was nearly an inch and a half in diameter and over three feet long, with one end flat for tamping, the other pointed. It was Gage who usually did the tamping.

One September afternoon in 1848, as Gage's team was blasting through an area of countryside near the town of Cavendish, a dreadful accident occurred. Perhaps someone forgot to add the sand or was momentarily distracted, we cannot be sure, but whatever the cause, as Gage tamped down into the hole, the iron rod sparked off the rock, igniting the powder and driving the tamping iron out – like a bullet from a gun – at tremendous speed. It entered 25-year-old Gage's head just below the left cheekbone, travelling upwards. Severing the optic nerve but missing the eye, it continued its destructive path through his brain, ablating portions of the left frontal lobe before exiting through the top of his head (Figure 5.1). It was said that the rod landed over 100 feet away, although this may be exaggerated.

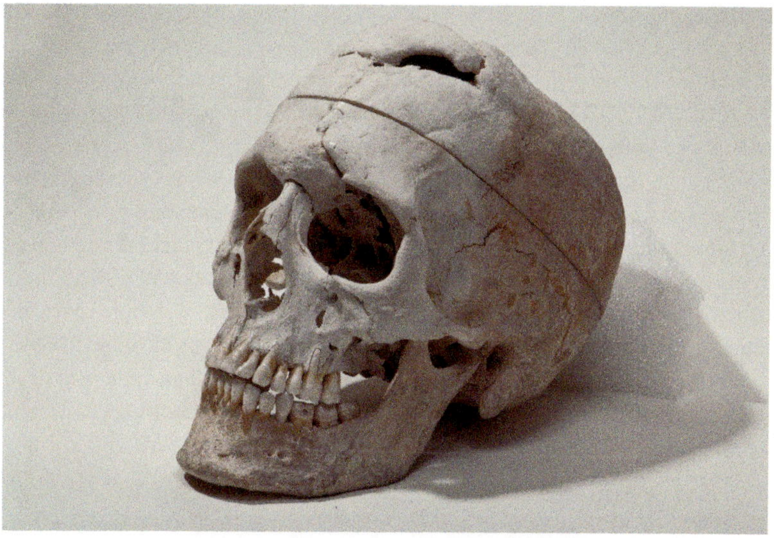

FIGURE 5.1 Skull of Phineas Gage showing the damage caused by the tamping iron, courtesy of Warren Anatomical Museum in the Francis A Countway Library of Medicine.

Remarkably, this injury did not kill Gage. He was unwell for a long period afterwards, suffering from a severe infection immediately after the incident and teetering at death's door for several months. But, over time and with care from a dedicated physician, Dr John Harlow, he did eventually recover.

But the most remarkable thing about Gage was the dramatic change in his behaviour and personality after the drastic loss of frontal tissue. While it is difficult to distinguish fact from myth in this regard – stories about him becoming a drunk, a gambler and a violent, profane blackguard have definitely been fabricated – what we can be certain of is the fact that he was a changed man afterwards. He was now easily distracted, found it difficult to focus his attention or organise his time, and had difficulty in breaking off from an activity he was engaged in. For example, if slicing a loaf of bread, he might continue to cut through the bread board (this is known as perseveration, a common feature of people who have suffered a frontal brain injury). Here was quite compelling evidence of the functions that were carried out by the areas of cortex obliterated in Gage's brain; the frontal lobes, it seemed, were responsible for attention, engagement in tasks and organising behaviour.

Gage lived for another 12 years after the injury. It was said that he spent some time in PT Barnum's museum, appearing for the public with his tamping iron (which he kept with him for the rest of his days) as a curio. He died in 1860 in San Francisco following a series of severe convulsions. And there the story appeared to end; his legacy was to endure ongoing fame in textbooks on medicine and neuroscience, psychology and cognition, as scientists pointed to the strange case of Phineas Gage as one of the earliest clues to the functions of the brain and as an example of how we can learn so much about the intact brain from the study of the injured one.

Generations of students were educated in the sciences knowing his name and fate, but none ever knew his face. The particulars of his unique story were depicted many times, in a variety of ways, ranging from early drawings of his accident to diagrammatic reconstructions of the path of the tamping iron through his head, photographs of his skull and the rod, death masks cast after his passing and – with the advent of technological graphic techniques – three-dimensional renderings of his tissue loss (Figure 5.2).

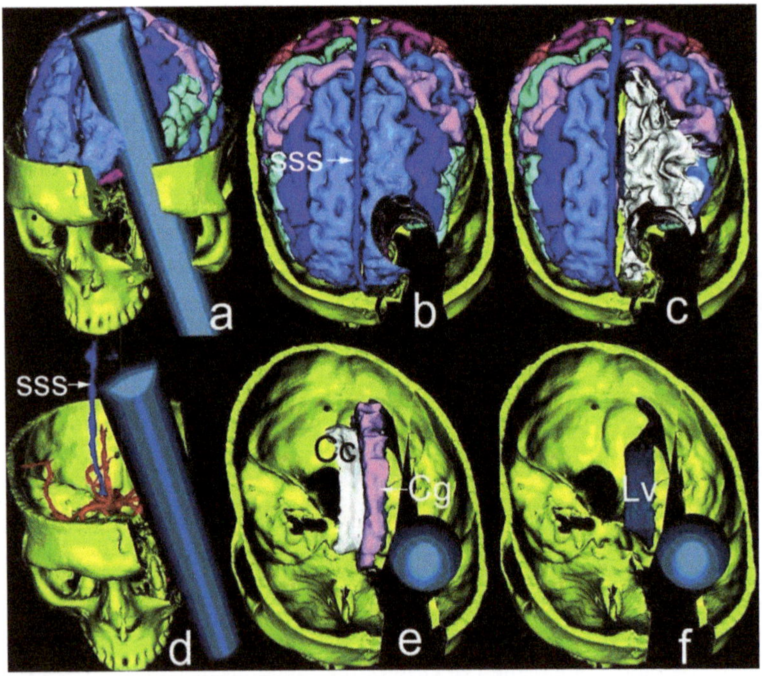

FIGURE 5.2 3D reconstruction of damage to Gage's skull with tamping iron. Image taken with permission from Ratiu et al. (2004). 'The Tale of Phineas Gage, Digitally Remastered', *Journal of Neurotrauma*, 21(5): pp. 637–643. The publisher for this copyrighted material is Mary Ann Liebert, Inc.

But each of these images retained a cold, clinical quality, as if showing the effects of some fictional event happening to an abstract, disembodied skull. Without a face to remind people that he was merely a normal man stricken by a tragic accident, Gage seemed an almost theoretical figure, his very humanity another aspect of his life lost on that afternoon in a flash of gunpowder. Just another brain with a hole in it. And that, it appeared, was how he was destined to remain.

That is, until 2008. In that year, Jack and Beverly Wilgus, a couple who ran an antique photograph company in the US, posted on their website an image of a well-dressed and handsome man with a sewn-shut left eye and holding a long pole (Figure 5.3, left). They called it *The Whaler* in the belief that the rod was a whaling harpoon, and that the facial scars the man displayed were linked to a violent encounter at sea. People soon contacted the Wilgus website informing them that the pole was not a harpoon, and some wondered if the man might possibly be Gage. They compared the man's face to the life-mask of Gage; the scars matched. They compared the rod to the tamping

FIGURE 5.3 Left: Daguerreotype of Phineas P Gage holding his well-known tamping iron (Warren Anatomical Museum in the Francis A Countway Library of Medicine. Gift of Jack and Beverly Wilgus). Permission received. Right: Second photograph of Gage from the Gage family of Texas photo collection. An identical image is in the possession of Phyllis Gage Hartley of New Jersey. (Author of underlying work unknown. File:PhineasPGage.jpg, Public Domain.)

iron, which is on display in the medical museum that also has Gage's skull; the inscription also matched. It reads:

> This is the bar that was shot through the head of Mr Phineas P. Gage at Cavendish, Vermont, Sept. 13, 1848. He fully recovered from the injury & deposited this bar in the Museum of the Medical College of Harvard University. Phineas P. Gage Lebanon Grafton Cy N-H Jan 6 1850.

Finally, after 160 years, the face of Phineas Gage, the man who inadvertently paved the way for much of modern neuroscience, was revealed to the public. It was widely believed that this was the only existing image of Gage following the accident, but within two months, a distant relative of Gage's brother, Tara Gage-Miller, produced a pocket portrait of the same man, whom she knew as a strange historical relative (Figure 5.3, right). She was aware of the story of his injury, but had no conception of the massive public interest in her distant uncle.

This happy chance revealed more than simply Gage's face – the serendipitous snapshot, taken by an unknown photographer oblivious to the future importance of his subject, would also go some way towards dispelling the myths about Gage's post-accident fate. There sits a clean-shaven, well-groomed gentleman, bolt upright in his chair, his tamping iron grasped proudly in hand. Not the aggressive drunk, not the gambler or layabout of rumour. This striking and somewhat melancholy image revealed not only the truth about Gage after his ordeal, but also the bitter reality of any brain injury survivor: that they are the same person. Only different.

The corners of my mind

For the past 60 years, all neuroscience students have – at one time or another – been told the story of Patient HM. Patient HM was an unremarkable young man growing up in East Hartford, Connecticut, in the late 1920s and early 1930s. He had had a stable childhood and was mannerly and pleasant, if a little shy. From the age of ten, he had begun suffering from *petit mal* epileptic seizures, brief episodes during which he appeared to drift off or become distracted from what he was doing. These continued for some years, and at the age of fifteen, he began to experience the more violent *grand mal* seizures more typically associated with epilepsy – convulsions, muscle contractions, shaking and loss of consciousness. These *grand mal* seizures persisted, and there followed a period of eleven years during which a succession of medications and treatments were employed – unsuccessfully – in an attempt to control his condition. So frequent and severe were these episodes that it proved impossible for him to hold down his job as a motor repairist. Finally, in August 1953, while under the care of psychosurgeon William Beecher Scoville, it was decided that HM should undergo a surgical procedure to remove the region of cortical tissue which was thought to be responsible for his illness. HM was 27 at the time, and the procedure was to take place at Hartford Hospital, with Scoville famously describing the surgery as 'a frankly experimental operation'.

Scoville had achieved some success in the past by using psychosurgery to alleviate psychosis, specifically by removing sections of the frontal or medial temporal lobes in sufferers, and he hypothesised that with HM, the source of his seizures lay in the same region of the brain. On 26th August 1953, two small holes were drilled just above each of HM's eyes; a thin implement was inserted and slid beneath the overhanging protrusion of the frontal cortex, and by applying suction, regions deep within his brain (specifically, the amygdala, parahippocampal gyrus and a large extent of his hippocampus) were removed from each cerebral hemisphere, an area of approximately 8 cm of tissue.

To Scoville's credit, the operation could be considered a success in that it did greatly reduce HM's seizures. But it very quickly became apparent in the days that followed the surgery that it had also left a catastrophic legacy on his experience of the world. Scoville began to notice that the young man no longer recognised the medical staff and was unable to make his way around the hospital. Further, he could not recall the day-to-day events of hospital life; he seemed to have completely lost his memory. Upon eating lunch on any given day, he would be unable to recall a single item of what he had eaten a mere 30 minutes before, or – worse still – that he had even eaten. He constantly had to be reminded where objects were located; his mother needed to tell him where the lawnmower was kept even though he may have used it the day before. He would read the same books over and over again without ever recalling that he had read them before, finding them fresh and novel each time. And the same was true for events that did not personally concern him – world events, too, were lost within moments of HM's learning of them. He had become a man frozen in the present moment.

This devastating condition is termed severe anterograde amnesia, an inability to retain any new factual information in memory beyond a very short period of time, in HM's case approximately 30 seconds unless he paid special attention. His IQ was unaffected by the operation, his ability to recall events from before his surgery was largely intact, and he could acquire new skills or motor patterns, revealing a distinction between personal (episodic) and procedural or implicit memory. HM's difficulty was a very specific one whereby he was unable to transfer new incoming information from his intact short-term memory to his long-term store. Every experience he had, every person he met, every new event faded into oblivion after the passage of mere seconds.

Four years later, in 1957, Scoville published a research paper with Brenda Milner (who was at the time a graduate student of the pioneering neurosurgeon Wilder Penfield) describing HM's procedure and the strange cognitive sequelae that had resulted. It was in this article that he was first referred to as Patient HM, the moniker that would be attached to him in the literature for the rest of his life, its brevity seeming to echo the tragically short-lived nature of his conscious experience.

For the next five decades, HM would become the most studied human in the history of neuroscience; the former motor repairist, by virtue of Scoville's tragic miscalculation, came to shed light on the workings of nature's most complex machine. HM spent those years living his disjointed life from moment to moment, his fleeting existence punctuated by regular visits to research centres and hospitals for cognitive assessments, test batteries, health checks and brain scans. The succession of major world events that occurred in those decades – from the Vietnam War and the assassination of Kennedy to the fall of communism and attacks on the twin towers – all briefly registered and were soon forgotten in the conundrum of HM's brain, as ethereal and short-lived as a dream. In fact, when once asked about his own experience of his condition, he stated that it was '… like waking from a dream … every day is alone in itself …'[1]

For the scientists who worked with him throughout that period – Brenda Milner, Suzanne Corkin and many others – he would become a large part of their lives, an affable, pleasant old man who took part in their experiments with good-natured acquiescence without ever remembering meeting them before. From these studies, much has been learned about the nature of memory, the stages of encoding, retrieval and storage, the process of consolidation, the differences between episodic, semantic (fact-based) and procedural memories, and the roles played by specific brain regions in these processes. In many ways, as much was gleaned from the things that HM could still do as from those he could not. As with Phineas Gage, HM's case also provided insight into the nature of identity and personality, aspects of ourselves so fundamentally linked to our memories. Despite his memory deficits, HM retained his sense of humour[2] and had a sense of right and wrong, with a strong sense of moral judgement; he never sought to blame Scoville for the damage his operation caused, as is evidenced in his interviews:

> Interviewer: Are you ever sad?
>
> HM: No, I feel what is done about me helps the doctors and the nurses, and everybody helps other people.

Interviewer: What a wonderful outlook.
HM: And I feel that's more important in a way. Because what they learn about me, they can also do on someone else, or not do on someone else.

Extract from an interview with Suzanne Corkin at MIT (1992)

The question of his self-image was also of great interest; did this now-elderly man whose memories ceased to accumulate in 1953 still view himself as a 27-year-old? When Corkin asked him if he could identify a photograph of himself and his mother, he replied that the man in the picture was his father. However, after a few moments of deliberation, he reconsidered and suggested that it could not be his father as he didn't wear glasses. While there was no immediate recognition of himself in the photograph, his initial assertion that the man in the picture was his father (perhaps due to recognisable familial traits and/or the presence of his mother) had to be rejected when a vital feature – the glasses – did not fit with this hypothesis. This ability to use different cues to bring about the correct answer is often observed in other cases of people with amnesia. Similarly, when HM looked at himself in the mirror and Corkin asked him what he saw, he did not express his horror at the old man he found staring back at him, but calmly stated 'I'm not a boy' (see Figure 5.4, left, for a photograph of HM taken in 1975).

Like Gage before him, Patient HM was, for many years, among the most famous cases in modern medicine, well-known by the many scans, diagrams and renderings of his surgery and postoperative brain (Figure 5.4, right), easily identifiable by the dark cavities where his hippocampi once resided. Throughout this time, he remained a celebrity without a face, more recognisable by his cortex than his countenance. This changed when, on 2nd December 2008, HM died of respiratory failure in a healthcare centre in Bickford, Connecticut (a nursing home founded by the brother of the nurse Lillian Herrick, with whom Henry had lived during the later part of his life). He was 82. Within hours of his passing, his brain had been scanned, harvested, preserved and cut into 2,401 incredibly thin slices for future study; these will be digitised and rendered into a full three-dimensional reconstruction, meaning that his contribution to science will continue even after his death.

This was a process long planned and ultimately overseen by the late Professor Suzanne Corkin, an MIT researcher who had worked with HM since 1962 and who had taken great care to limit exposure to him and protect his anonymity[3] throughout her long association and friendship with the man who could never remember her. HM's obituary was published in

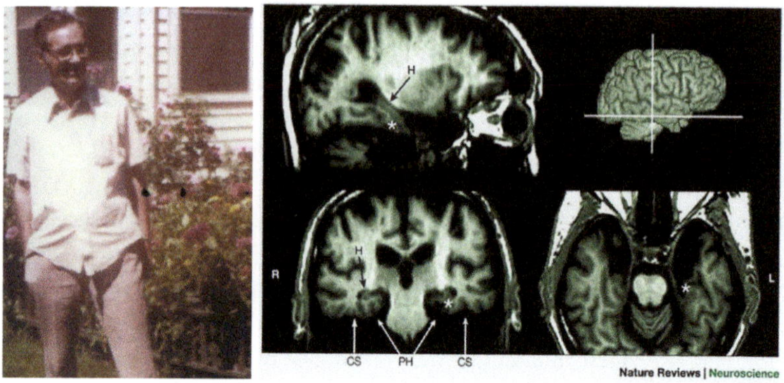

FIGURE 5.4 Left: Photograph of HM taken in 1975 (image from *Permanent Present Tense* by Suzanne Corkin. Copyright Suzanne Corkin, 2013, used by permission of The Wylie Agency (UK) Ltd. Right: MRI scan of HM's brain depicting the loss of tissue (Abbreviations: CS, collateral sulcus; EC, entorhinal cortex; H, hippocampus; L, left; PH, parahippocampal gyrus). Reprinted by permission from Macmillan Publishers Ltd: (Corkin S,'What's new with the amnesic patient H.M.?', *Nature Reviews Neuroscience*, February 2002; 3[2]:153–60).

The New York Times two days later, where his identity was finally revealed to the world as Henry Gustav Molaison; the world's most famous amnesiac.

> Henry Gustav Molaison, born on Feb. 26, 1926, left no survivors. He left a legacy in science that cannot be erased.
>
> *Carey (2008)*

We do not know if Henry dreamed; if he did, it is likely that, as with most of the experiences that briefly danced through the connections of his wounded brain, his dreams were quickly lost again to the darkness. But we can hope that, in the forgotten dream times of his sleep, the man who compared his waking life to emerging from a dream might, in those moments, have felt fully awake, fully alive.

A shrewd awakening

The year 2008 was a significant one for neuroscience. Not because of any specific publication, research finding or ground-breaking discovery in the field; as with any year, there were many such advances in 2008. The reason 2008 will stand out as a landmark date in the history of neuroscience is that it was the year in which, for the first time, we were allowed to put a face to two of the world's most famous brains.

Out of remarkable serendipity, new generations of students will now learn the story of Phineas Gage, but he will no longer be an abstract, vague spectre of a nineteenth-century railwayman. And thanks to the devotion of those who worked with him and were his friends, we will read of Patient HM and know that he was Henry Molaison, and attach a face to that new name – something he could never do. We can at last look on the faces of two proud men – the owners of science's most famous brains – who each had so much taken from them, but who gave back more than they could ever have imagined.

The brain loves to make connections; this is probably its primary reason for existing. It is perhaps fitting, then, that these two remarkable men should be forever connected, not only due to their unique medical histories and what was learned from them, but also by the common year in which the world finally got to meet them. Even after their passing, they are an important reminder of the lives – the sometimes shattered lives – that lie within every medical case study and set of initials; a reminder that they are people, not walking collections of preserved and lost functions. This is a message beautifully expressed in the writings of the great British neurologist and author Oliver Sacks.

Oliver Wolf Sacks (1933–2015) began his career in the 1960s working with patients suffering from neurological damage when, following his medical degree at Oxford and a time at Middlesex Hospital in London, he moved to Mount Zion Hospital in San Francisco. His early interest in chemistry and, later, biology had set him on an inexorable path towards medicine, psychopharmacology and, ultimately, neurology, the field with which he would become synonymous. Here, while working at Mount Zion and conducting research at UCLA, he was exposed to a wide variety of rare and fascinating patients, people with discrete or unusual cases of brain injury or impairment, cases from whom the medical community could learn much. This continued when Sacks moved to the east coast in 1965, taking up a post at Albert Einstein College of Medicine in New York and a year later as neurologist at Beth Abraham Hospital in the Bronx.

For the rest of his life, Sacks continued to work with and document the strange cases which he encountered, acting as neurological consultant to several institutions in the New York area. During a prolific career, he produced nine published volumes of collected case studies[4], most famously *Awakenings* in 1973, later turned into a Hollywood movie, and *The*

Man Who Mistook His Wife for a Hat in 1985, which has been converted into both a stage play and a musical. Yet it was not for the scientific merit of these collections – which was high – that he is best remembered, but the staggering compassion with which he recounted his patients' stories. The combination of his remarkable writing with his ability to retain the humanity of his charges led the *New York Times* to call him 'the poet laureate of contemporary medicine'.

Sacks' great gift was his ability to remember, and convey to the reader, that what sat before him in each clinic was not merely a collection of symptoms and deficits but a human being, a person robbed of a part of themselves through accident or disease. He was probably influenced in this by the writings of the poet WH Auden; Sacks had admired his work in his youth, and the pair had later become friends. While writing *Awakenings*, Auden urged him to move beyond mere clinical description. Sacks, it seems, took this advice to heart. The recurring chord which resonates through his writings, from *Migraine* to *An Anthropologist on Mars*, from *The Island of the Colourblind* to *Musicophilia*, is his deep and profound identification with each person, their sense of loss and struggles to accept and adapt to the fates which befell them. In each case, he outlines the neurological basis of the problem and beautifully describes the effects of the damage in lyrical and imaginative detail, but never loses sight of the person playing host to the condition. As readers, he makes us *feel* the awkwardness of The Disembodied Woman and the bewilderment of The Lost Mariner. In that ability, that kindness, lay Sacks' true genius. In his own words:

> If we wish to know about a man, we ask 'what is his story – his real, inmost story?' – for each of us is a biography, a story. Each of us is a singular narrative, which is constructed, continually, unconsciously, by, through, and in us – through our perceptions, our feelings, our thoughts, our actions; and, not least, our discourse, our spoken narrations. Biologically, physiologically, we are not so different from each other; historically, as narratives – we are each of us unique.
>
> *Sacks (2009, p. 110)*

The eclectic and remarkable contents of his own experience also clearly informed his outlook on his patients and their situations; the former weight-lifter, renowned swimmer and motorcycle enthusiast lived his own life with a deep spirituality, a powerful belief in the transformative power of music, and a pragmatic and gentle acceptance of mortality. His own experience of face blindness (prosopagnosia) clearly informed his description of 'Dr P', the prosopagnosic patient who gives *The Man Who Mistook His Wife for a Hat* its name. In his 1984 memoir, Sacks recounts the tale of his brush with a near-fatal accident when he seriously damaged his left leg while escaping from a bull in a Norwegian fjord. As he attempted to crawl and scoot his way down the mountainside before the fall of night, he used music to dictate the tempo of his body movements, lapsing into a rhythmic, almost trance-like state of automatic action.

Oliver Sacks – the man who for fifty years worked with people and their broken instruments, whose words made their music audible again – died in August 2015 at the age of 82. The profound humanity and humbling compassion of his writings leave us with an important and timely legacy: to remember that behind the disease, case study or data point lies a person, a human being grappling with their unique and often distressing reality.

> One must drop all presuppositions and dogmas and rules – for these only lead to stalemate or disaster; one must cease to regard all patients as replicas, and honor each one with individual reactions and propensities; and, in this way, with the patient as one's equal, one's co-explorer, not one's puppet, one may find therapeutic ways which are better than other ways, tactics which can be modified as occasion requires.
>
> *Sacks (2010, p. 219)*

Notes

1 See Milner, Corkin and Teuber (1968).
2 When Corkin once asked him what he did in order to attempt to remember material, he replied, 'Well, that I don't know 'cause I don't remember (laugh) what I tried'.
3 Since his death, several photographs, audio recordings and interview transcripts of HM have been released into the public domain.
4 The final count would have been ten had he not burned his first attempted collection, *Ward 23*, in 1967 in a fierce crisis of self-doubt.

References

Carey, B. 2008. H. M., An Unforgettable Amnesiac, Dies at 82. *The New York Times* (4 December).

Milner, B., Corkin, S., and Teuber, H.L. 1968. Further Analysis of the Hippocampal Amnesic Syndrome: 14-Year Follow-up Study of HM. *Neuropsychologia*. 6(3), pp. 215–234.

Sacks, O. 2009. *The Man Who Mistook His Wife for a Hat*. London: Picador.

Sacks, O. 2010. *Awakenings*. London: Picador.

6

IMAGING ART PERCEPTION

Introduction

In the first section of this book we considered the historical links between science and visual aesthetics, chronicling the many ways in which pictorial, and frequently artistic, representations of scientific ideas, observations and data were used to powerful effect in explaining, simplifying and disseminating novel concepts. This section explores contemporary views on visual art, science and the brain. We first outline what is currently understood about the cortical systems responsible for visual perception (this chapter), describing the different pathways and processing streams which allow us to perceive the visual world around us. In Chapter 7, we consider the symbiotic relationship between the brain and visual art, showing first how neurological conditions have occasionally been revealed through the output of artists, and then relating unusual cases of how brain injury has led to a change in style among artists or the emergence of *de novo* artistic ability in non-artists. Finally, in Chapter 8, we discuss the newly established field of *neuroaesthetics*, the formal study of the brain's response to the perception and appreciation of (mainly visual) art, describing the different strands of investigation within this new discipline and summarising the key debates and issues surrounding this controversial subject matter.

This chapter deals with the nature of the human visual system, the biological substrates of our ability to see and recognise the contents of our visual world. But rather than provide a standard textbook description of the information-processing pathways from retina to occipital cortex and through the various specialised brain regions thereafter, we will convey the working of the visual system by means of analogy, by recounting the remarkable and little-known story of – appropriately – an artist, Johannes Matthäus Koelz.

Missing links: Koelz and *Du Sollst Nicht Töten*

We humans find something almost irresistible about an incomplete work, in the suspense of a mystery story with the final page torn out. In fact, evidence seems to suggest that our brains spend much of their time in a state of waiting for the other shoe to drop. The short biography that follows tells the story of a German artist, Johannes Matthäus Koelz (1895–1971). His life was a turbulent one, lived in dangerous times; it was filled with drama and tragedy, serendipity and triumph. Yet through the recounting of his remarkable tale, we can discover unusual truths about the nature of that humanity and how it arises, about the elegant and ingenious ways in which our brains give rise to what makes us what we are.

In 1915, 20-year-old Koelz was sent to the trenches of the Western Front as an infantry officer. Despite receiving the Iron Cross for courage for rescuing a fellow soldier from a collapsed trench at Verdun, Koelz's experiences of the inhumane conditions on the Front had a profound impact on his views on warfare and nationalism. Following the end of World War I in 1918, he spent five years serving in the police forces of Munich, Bavaria and later Berlin. His growing sense of disenchantment with the establishment appears to have been galvanised during this time, and by 1924 Koelz had fully adopted the principles of pacifism, becoming an outspoken member of the growing German Pacifist movement. Also at this time, Koelz was completing his extensive studies at the Academy of Fine Arts in Munich – a twelve-year apprenticeship. As a master student of Franz von Stuck and Olaf Gulbranson, Koelz developed into an exceptional painter in the realist tradition of Hodler and Leibl. His reputation steadily grew after gaining a few commissions, and his artistic prowess developed in parallel with his pacifist activities. It was the combination of these two interests that would later define his life and the strange direction it took.

Koelz began work on his masterpiece, a large triptych (8 x 24 feet) called *Du Sollst Nicht Töten* (Thou Shalt Not Kill!), in 1924. It was to be his great anti-war propagandist statement: a huge depiction of a crucified German soldier with decomposing Allied troops at his feet in the central panel, flanked by images of soldiers, civilians, children and church leaders in each side panel, symbolically lending their support to the slaughter of World War I. The triptych travelled with Koelz throughout the decade that followed as he moved from Munich to Slovenia to Bavaria and back to Munich again.

FIGURE 6.1 Johannes Matthäus Koelz, *Du Sollst Nicht Töten* (Thou Shalt Not Kill!), Leicester Arts and Museums Service (New Walk Museum and Art Gallery). © Estate of Johannes Matthäus Koelz, DACS London/IVARO Dublin, 2017.

Koelz secretly worked on the piece from 1924 to 1937 against a backdrop of increasing political and social uncertainty, all the while continuing his pacifist propaganda through leaflets, poetry and other media. As the atmosphere in Germany became steadily bleaker, it seemed to Koelz that both his activism and his work on the triptych grew increasingly urgent.

In 1933 Koelz took a photograph of the piece (Figure 6.1). It was still unfinished at this time: only one of the rotting Allied soldiers had been painted, some figures were missing a hand, portions of the background remained blank. This solitary black-and-white photograph was to be the only lasting record of the triptych as a coherent piece. For, four years later and with his master work finally complete, Koelz's anti-Nazi activism caught up with him. Earlier in 1937 he had received a commission to paint a portrait of Hitler, such was his profile in the German art world at that time. His refusal had brought his propagandist activities to the notice of the secret state police and – charged with 'insulting the Führer and spreading pacifist propaganda' – a warrant was issued for his arrest.

Then a truly remarkable thing happened. In an event of most unlikely serendipity, the officer despatched to arrest Koelz was the same soldier whose life Koelz had saved in the trenches at Verdun. He granted the artist a chance of escape, allowing him 48 hours to leave Munich with his young family. This act of kindness presented Koelz with an agonising dilemma – to flee Germany and leave the enormous painting behind to certain destruction as anti-Nazi propaganda, or to face arrest himself and condemn his opus to the same fate.

Koelz hit upon a solution that was as traumatic as it was ingenious; he decided that to save his art, he must take it apart. He brought the giant triptych to a nearby saw mill and instructed the young man working there to cut the painting into more than twenty smaller pieces – a face, a soldier, some flowers, a pair of hands joined in prayer. Each of these fragments was to be hidden, passed on to trusted friends or relatives, kept secret until a day might come when it would be safe to reunite them. It is difficult to imagine the crushing mixture of emotions that must have converged on Koelz as he gave up his life's work, his epic statement of what he believed, to the teeth of the sawblade.

The rest of Koelz's life, like his triptych, is fragmented and scattered. He fled from Munich to Austria, then to Prague, and later to England in 1939. He was interned as an enemy alien in 1940 and sent, along with nearly 3,000 other survivors who had escaped from Nazi Germany, to the deplorable conditions of Hay Camp in New South Wales aboard the SS Dunera (narrowly avoiding being sunk by torpedo). He returned to England a year later, and eventually to Germany after the war. Following the death of his wife Claire in 1957, he divided his time between Stoke-on-Trent and Curraglass, county Cork, painting and sculpting until his death in 1971.

And there his story should end. But it does not, because now, over 70 years later, the separate pieces of his triptych are gradually being brought back together. Working from the grainy black-and-white photograph from 1933, the surviving friends and relatives of Koelz are scouring the galleries, museums and private collections of Europe in an attempt to uncover the lost pieces of the puzzle. To date, six have been found, while rumours circulate of some of the other paintings being sighted in collections in Poland and Austria. The recovered pieces are displayed in exhibitions against the enlarged monochrome of the photograph; three from the left panel, one from the central, two from the right. It seems as though Koelz's inspired if harrowing solution might finally succeed – that one day in the distant future, the final missing piece of his epic work will be returned to where it belongs and the shocking and powerful message of *Du Sollst Nicht Töten* will be restored and renewed[1].

Parallels: Koelz and the visual system

A striking similarity exists between the fate of Koelz's masterpiece and how visual information is processed in the brain. The way in which the triptych was taken apart and separated only to be later reconstructed (or so he hoped) bears a startling resemblance to the way in which any visual scene – a landscape, a face, a painting – is broken down by our visual system, the elements kept separate, only to be recombined at a later stage. Decades of research into the visual processing areas of the occipital lobes and beyond have revealed that visual images falling onto the retina are fragmented at a very early stage of processing, with different aspects of the scene processed in functionally separate and separable pathways from retina to lateral geniculate nucleus and onward to primary and secondary areas of visual cortex[2].

Figure 6.2 provides a summary of how visual information is segregated and processed along parallel pathways from the eyes to the primary visual

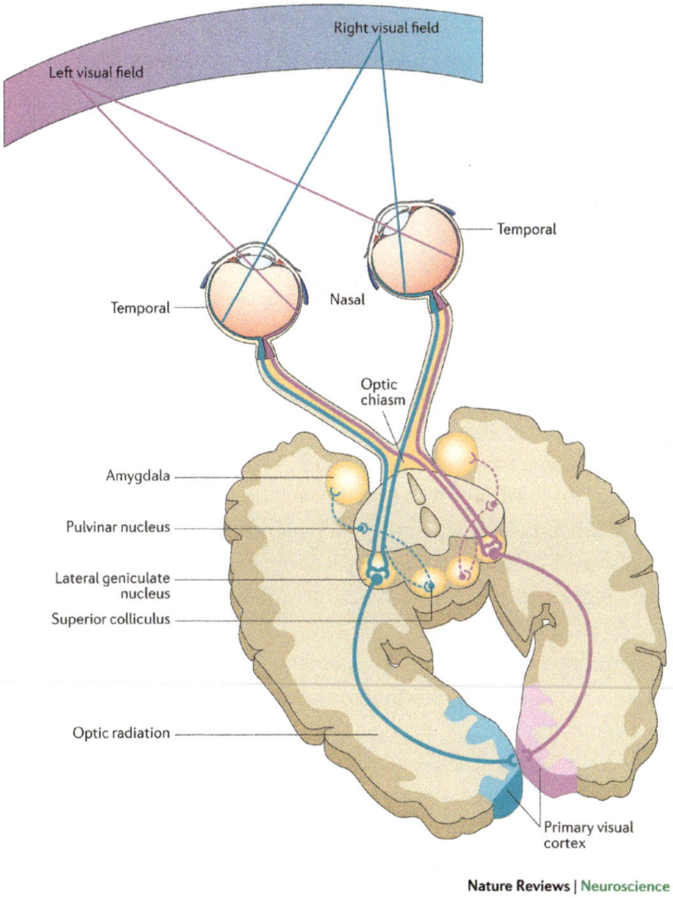

FIGURE 6.2 Overview of the human visual system. Reprinted by permission from Macmillan Publishers Ltd: *Nature Reviews Neuroscience*, Hannula et al. (2005). 'Imaging implicit perception: promise and pitfalls', *Nature Reviews Neuroscience*, 6(3), pp. 247–255.

cortex, located at the back of the brain. The scene we observe is initially flipped upside down by the lens in our eyes; any information in our right visual field is projected to the left part of the retina (blue lines), while information from the left visual field is projected to the right (purple). This segregation continues along the optic nerve through to the brain so that all information pertaining to the right visual field ends up in the left primary visual cortex while information related to the left visual field is processed by the right visual cortex.

David Hubel and Torsten Wiesel painstakingly recorded the electrical firing properties of neurons in the primary visual cortex in response to light. They discovered that single neurons only responded when light was presented in a specific region of space, and that visual space was mapped topographically onto the cortex. More than this, they found some cells that only responded to light at a particular orientation, other cells that responded only to the movement of light, and yet others that responded to the wavelength of light; that is, colour. In this way, information pertaining to the full visual scene and details regarding colour, movement and fine detail are dealt with by separate regions and even by separate neurons within the primary visual cortex (area V1). This segregation is maintained into the next cortical region (area V2).

Beyond V2, this separation of function branches out further to distinct cortical regions: area V4 is specialised for colour processing, V5 (also known as MT) for movement, parts of V3 for form (V3a) and dynamic form (dV3). Later still, information on an object's identity and where it is located are handled in separate, parallel streams, the so-called 'what and where pathways' of the temporal and parietal lobes respectively. Yet despite these segregations and separations, the different areas remain exquisitely interconnected, a dense network of cross-talking circuits allowing all the discretely processed aspects and fragments of the scene to be brought back together and reconstituted into the complete and meaningful percept of an object, a person, a scene that occupies our conscious awareness. Like Koelz's masterwork, the individual pieces are scattered to the four corners of the brain only to be reunited and made whole once more, albeit in a much, much shorter timescale.

The visual system also holds another phenomenon that is hinted at by the Koelz story, although it is a quirk of the human brain that is not limited to vision. The phenomenon is that of the *scotoma*, a small blind spot in the visual field which results from minor damage to the primary or secondary visual cortex. Due to the retinotopic arrangement of cells in V1 and V2 – whereby adjacent neurons process information from adjacent areas of the retina and, therefore, the visual scene – a person suffering from a scotoma is left with a small hole in their visual world. Remarkably, however, a sufferer may not even be aware of the presence of a scotoma due to the brain's tendency to fill in blanks, just as it does with the blind spot which we all possess. Our brains are constantly playing detective, compensating for gaps in perception with a 'best guess'. As the psychologists of the Gestalt school discovered, our visual systems seem predisposed to complete the circle, continue the line, or group stimuli according to such features as similarity. These perceptual acts are carried out unbidden, without our control or conscious awareness. They also hint at a deeper principle about how we process information: just as nature abhors a vacuum, our brains appear to abhor incompleteness – there is a drive to finish the unfinished, to put the last jigsaw piece in place. This may explain some of the fascination with Koelz's uncanny story, and it has even been proposed as one of the bases of Neuroaesthetics, an emerging field which attempts to identify the mechanisms underlying our ability to appreciate aesthetic beauty (discussed in Chapter 8).

The remarkable life of Johannes Matthäus Koelz and the tantalising puzzle he left behind tell us much about the resilience of the human spirit, about its ability to endure the most appalling and profound hardships, both physical and emotional. But it may also afford us some fleeting glimpses of the subtle mechanisms behind this inspiring and humbling human condition, if only by means of metaphor. The story of Koelz's triptych is as yet incomplete – all the pieces may never be found (Figure 6.3). Likewise, our understanding of how we process information, how we perceive, how we appreciate a thing of beauty, still has far to go. In science, as in art, it would seem that we are still waiting to learn how the story ends.

FIGURE 6.3 Artist's impression of what the completed tryptych would look like (art: J Koelz, art restoration: Daryl Joyce).

Disorders of visual perception

Before any interpretation or evaluation of a work of visual art can take place, a succession of processing stages within the visual system must first take place to allow the perception, identification and recognition of the piece to be accomplished. Our ability to experience visual art is therefore dependent on our ability to perceive it. In addition to scotomas, which leave a gap or hole in the viewer's visual experience, there are a number of extremely specific visual disorders which can arise as a result of discrete damage to specific regions of the occipital cortex and the visual areas beyond. Due to the functional specialisation of these cortical regions, the impairments do not affect the visual scene as a whole (as in a scotoma), but rather distinct aspects of it. It is to these disorders that we next turn our attention.

After visual information passes from occipital area V2, which processes all aspects of a scene, to more specialised regions such as V3, V4 and V5, the retinotopy of primary and secondary visual cortices is abandoned, leading to a collection of more specific visual processing areas. Damage to regions specialised for the processing of shape or form, such as occipital area V3a or temporal visual areas like TE or TEO (which together make up the ventral, object-processing or 'what' pathway) can cause an inability to identify or recognise visual objects. Such *visual object agnosia*, which can occur as a result of a gunshot wound or a penetrative head injury, is a specific dysfunction of visual access to the correct term; an agnosia sufferer might be unable to access the term 'rose' when shown an image of the flower, but may be able to name it normally when allowed to smell or touch the object. A more extreme form of this, *prosopagnosia*, involves an impairment in the ability to recognise faces, and results from damage or dysfunction in the fusiform face area (FFA), another ventral temporal region. To a prosopagnosic, the features of faces – even familiar ones including their own – appear jumbled, confused or incoherent. The late Oliver Sacks – himself a sufferer of the disorder – most famously described the case of Dr P, the prosopagnosic whose experience was coined in the title of his book, *The Man Who Mistook His Wife for a Hat*. Other notable cases include primatologist Jane Goodall, comedian Stephen Fry and actor Brad Pitt.

Disturbances to the other specialised visual areas can result in similarly specific deficits. Damage to area V4, where cells are particularly responsive to processing colour, can result in an inability to perceive in normal colour vision, *cerebral achromatopsia*. Patients with this condition are confronted with a visual world in black and white, although – due to the fact that bilateral damage to V4 is rare – normally only one side of their visual field is achromatic. Similarly, injury to area V5 (part of the dorsal 'where' pathway), which is functionally specialised for motion processing, results in *cerebral akinetopsia*, wherein one is unable to perceive movement in the visual field. Such people are left to experience the world as if through a series of snapshots or a strobe-lit scene, in the absence of the smoothly-flowing movements of objects in motion. Damage to further areas of the dorsal stream – including parietal lobe regions – can lead to *apraxia*, or visually guided misreaching, whereby patients over- or under-shoot their movements when reaching for a visual object (despite, in some cases, making the appropriate hand-shape to successfully pick up the item). These and other examples of cortically based visual impairments underscore the truth in the statement that we see with the eyes but we perceive with the brain.

In order to perceive the visual world that surrounds us – to comprehend and appreciate the vividness of colours, the complexity of shapes, the nuances of facial expressions, the fluidity of movement – we rely on an astonishingly sophisticated and exquisitely elegant system. This system is capable of processing the blotches of light of varying wavelengths which fall on our retina, transporting them from eye to brain, separating them out, interrogating them in parallel, breaking apart the visual image received by the eyes into its most basic components, dealing with the different aspects in separate functional areas, and eventually recombining them to form recognisable shapes – familiar objects, people, faces, moving images – all in less than the blink of an eye. So rapidly do these processes take place that we can be forgiven for feeling as though the act of seeing is instantaneous. As with many of the brain's systems, we often fail to notice that any process has taken place until something happens to render it dysfunctional. By observing the effects of damage to visual areas on the perceptual experience of the sufferer, we have learned much about the nature of visual processing. In the next chapter, we will address more directly the relationship between the brain and visual art.

Notes

1 Farrington (1995).

2 A second, and potentially older, visual processing pathway exists which bypasses the visual cortex and projects directly from retina to higher-order visual areas via tectum and pulvinar; this pathway, proposed by some to underpin the phenomenon of blindsight, will not be considered further here.

Reference

Farrington, A. M. 1995. *Three Point Perspective: A Biography of Johannes Matthäus Koelz, 1895–1971*. AMF Research and Publication: Heather, Leicestershire.

7

VISUAL ART AND THE BRAIN

Introduction

In Chapter 6 we described the mechanics of the primate visual system – from the eye, via the retina, to the brain – and described the intricate and elegant means by which incoming images are deconstructed, processed in parallel in specialised areas, and later recombined to allow the recognition of and interaction with the visual world around us. As perception is so often the lens through which visual aesthetics gazes, it is unsurprising that a close relationship exists between visual art and the brain systems responsible for vision. The historical record is dotted with cases of artistic output providing clues to the neurological state of the artist, often unintentionally. As mentioned in Chapter 4, early cave art includes representations of people, animals, constellations – but perhaps most intriguingly – abstract patterns of dots, lines and lattices, which some theorists suggest may be attempts to depict the experience of scintillating scotoma, the sparking dots that can appear in the visual field[1]. Visual art, therefore, can act as a window into the mental state of the artist, providing neuroscientists with a subtle glimpse of the workings of their brain. As the composer John Cage states:

> The function of Art is to imitate Nature in her manner of operation. Our understanding of her manner of operation changes according to advances in the sciences.
>
> *(1969)*

In this chapter, we first consider a series of instances of artists who suffered brain injuries and look at how the effects of these insults were manifested in their subsequent works. Next, we describe fascinating cases where similar brain trauma has led to a change in or even an enhancement of artistic output. We then consider remarkable examples of non-artists who have found their artistic abilities unlocked following neurological conditions. Finally, we explore individual idiosyncrasies, such as synaesthesia and tetrachromacy, which provide people with unique ways of viewing their visual world, and look at how these conditions can be represented in the artistic outputs of such individuals. We note that this is by no means a comprehensive review, and refer the reader to works by Chatterjee and colleagues[2] who discuss many other cases not covered here, including instances of artistic ability in OCD (e.g. Franco Magnani), subarachnoid haemorrhage, autism (e.g. Nadia), right hemisphere injury (e.g. Loring Hughes), and left hemisphere injury (e.g. Zlatio Boiyadjiev and Katherine Sherwood).

Art disrupted by brain injury

Many artistic techniques require the integration of many skills and abilities including visual, spatial and motor. These sensory-dependent processes are complemented by higher-order functions of planning, decision-making, abstraction and imagination. Given their reliance on so fragile an organ as the brain, the consequences of neurological injury to an artist can be catastrophic. In this section, we relate five cases of such brain damage and the resultant detriment to the artists' output.

One particularly poignant case, described in the neurological literature by Sacks and Wasserman, is that of the artist JI. In 1986, following a car accident, JI suffered a concussion and became colour blind. For an artist who had dealt in abstract colour paintings, this was a particularly devastating blow. Initially he described his visual world as foggy, as if a mist had descended upon him. Hoping that this fog would eventually lift, JI decided to return to his studio the day after the accident. However, instead of seeing his studio filled with the rich colours of his paintings, it was now grey and devoid of all colour.

Over the course of the following months he began to despair, feeling that his art, even his life, were without meaning. It was not the case that his colours were gone, but everything he saw in his environment was black, white and dirty grey. As the months and years passed, even his dreams began to lose colour until eventually he was unable to remember or even imagine colour. His friends encouraged him to continue painting and, after a number of desperate attempts, he switched from painting in colour to black and white. He found some hope in this new outlet for his artistic abilities. He also took up sculpting, tapping into other visual dimensions such as shape and movement, capacities that were unaffected by his condition.

Two years after his accident he described to Sacks how his life had changed completely – he had become a person of the night; he drove and wandered about at night, experiencing a certain comfort in the darkness after sunset. Adapting to his changed circumstances, JI found that he could live with less despair in his colourless world. Although tragic on a personal level for the artist, this story shone a light on the functioning of the colour-processing regions of the visual cortex, area V4 (see Chapter 6), revealing that it is involved not only in the perception of colour, but also in colour imagination and memory.

Other neurological cases show less sudden but equally devastating changes. There have been a number of well-documented cases of artists who have gradually developed Alzheimer's disease or similar dementias. By tracking the progressive changes in their artistic work, scientists can chart the visuospatial and cognitive deficits associated with the disease.

One such study describes the deterioration in the artistic output of William Utermohlen (1933–2007), a 66-year-old portrait artist who was diagnosed with Alzheimer's disease following a marked decline in his memory and other cognitive capacities. Using a selection of self-portraits, the authors of the study were able to track the progressive nature of the disease; a self-portrait produced at age 60, before he or his wife noticed any cognitive difficulties, reveals precise brushwork, strong vivid colours and excellent emotional expression (Figure 7.1a). By the age of 62, Utermohlen was beginning to experience cognitive decline and his self-portrait from this time shows signs of difficulty representing individual facial features, particularly in terms of their structure and spatial orientation (Figure 7.1b). Such difficulties are more noticeable two years later, when certain facial features – such as the ear – are completely out of proportion. There is also little or no background to the painting (Figure 7.1c). By the age of 64, facial features have become disjointed, absent or blurred (Figure 7.1d), reflecting the downward trajectory of cognition associated with this disease.

Similar artistic deterioration has been reported in other Alzheimer's cases, with Cummings and Zarit describing the work of an artist over a 30-month period as becoming thematically more simplistic, with less elaborate colouring and a gradual loss of shading and perspective. Other notable cases of artistic decline during Alzheimer's disease include that of Willem de Kooning (1904–1997), whose works show a steady decline which parallels

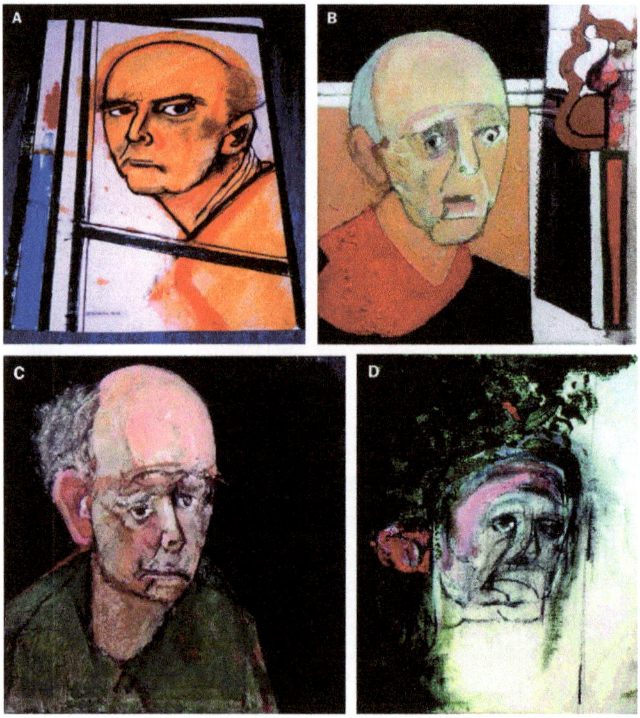

FIGURE 7.1 Self-portraits of William Utermohlen showing a deterioration in his artwork with the progression of Alzheimer's disease. Portrait painted at age of (a) 62, (b) 62, (c) 63 and (d) 64. Image taken from Crutch et al. (2001), 'Some workmen can blame their tools: artistic change in an individual with Alzheimer's disease', *Lancet*, 357(9274), pp. 2129–33 with permission from Elsevier.

the progression of his dementia to the point that his work eventually became disorganised to the point of meaninglessness.

In fact, recent work[3] has suggested that the presence of neurological disorders such as Parkinson's or Alzheimer's disease may be revealed in the works of painters long before a formal diagnosis takes place. The authors, Forsythe, Williams and Reilly, analysed a metric known as the fractal dimension (a measure of complexity) of 2,092 works of art completed by seven artists – two with Parkinson's disease (Dali and Morrisseau); two with Alzheimer's disease (Brooks and de Kooning), and three with no diagnosis (Chagall, Picasso and Monet). Using this metric, they could accurately predict the artists' subsequent diagnoses years prior to the actual clinical diagnosis. Thus, it appears that the ability of artistic output to reflect the presence of neurological dysfunction may be more than a case of *post hoc* anecdotes, with metrics such as fractal dimension potentially yielding predictive power.

Another common cognitive difficulty experienced by sufferers of Alzheimer-type dementias is a deterioration in the ability to recognise faces, while some may also experience mood and behavioural changes such as depression, delusions and hallucinations. All of these can have a serious impact on a person's everyday life, as well as their artistic output.

Maurer and Prvulovic conducted an in-depth analysis of the paintings of Carolus Horn (1921–1992), a German painter of note who was diagnosed with Alzheimer's disease in the mid-1980s. Similar to other patients, Horn showed a gradual confusion of perspective and a steady deterioration in the representation of spatial relations. This can be observed in his painting of the Bridge of Sighs in Venice in 1986 (Figure 7.2, right). When compared to his previous 1981 work (Figure 7.2, left), the later work shows less elaboration and impaired spatial relations. In the later work, the bridge in the background is much larger, the posts for the boats are more skewed and the water seems to stop suddenly compared to the greater realism in the 1981 painting. Two other major differences are visible: the cartoon-like representation of individuals and

FIGURE 7.2 Two paintings of the Bridge of Sighs, Venice, by Carolus Horn before (left) and after diagnosis (right) of Alzheimer's disease. (Image taken from Maurer and Prvulovic [2004], 'Paintings of an artist with Alzheimer's disease: visuoconstructural deficits during dementia', *Journal of Neural Transmission*, 111(3), pp. 235–45 with permission from Springer.)

the complete change of colour post-diagnosis. Such differences may hint at a lack of discrimination and the inability to recognise faces on the part of Horn. Furthermore, the greater preference for yellow–red colours may also reveal an attempt to compensate for this lack of discrimination; some research has shown that Alzheimer's patients show better discriminatory abilities in yellow, red and lighter colours in contrast to colours in the blue–green range, leading to suggested colour schemes for dementia care homes and clinical settings.

While neurological disorders like Alzheimer's disease are degenerative and progressive, patients suffering from other conditions, such as stroke, can recover a considerable degree of functionality. Through perseverance on the part of the patient and personalised long-term clinical rehabilitation programmes, patients may recover many of their lost abilities, including movement and language functioning. The brain can be a remarkably resilient organ; it had previously been thought that once the brain suffered damage there was little chance of recovery. However, it is now known that the brain is highly plastic, and that the neural pathways and circuits of the cerebral cortex can remap and rewire themselves in response to injury. In many cases, intact areas adjacent to the damaged region can take over the lost functionality. Through training and repeated stimulation, partial or full functionality can be restored, providing hope for many stroke survivors. The next two cases of artists who survived stroke damage illustrate the types of deficit which can occur from such damage, but also the staggering capacity for recovery possessed by the human brain.

Damage to specific brain regions during stroke (where the brain's blood supply is either cut off or a blood vessel ruptures) brings about some very specific and unusual deficits. Left-hemisphere damage often results in language difficulties, including impairments in producing and understanding speech. In contrast, right-hemisphere injury can lead to a number of visuo-spatial issues including left-sided hemineglect (sometimes called Unilateral Spatial Neglect), an inability to perceive or attend to the left side of the visual world. Hemineglect has been observed in many artists after stroke, resulting in blank spaces on the left side of their paintings and other artistic works[4]. Otto Dix (1891–1969), a German painter who depicted many naturalistic scenes of World War I and the Weimar Republic, suffered a right-hemisphere stroke in 1967. In the initial few days following his stroke Dix was unable to draw anything; by day four he began to draw a small sketch of a tree, but the left side of the picture remained blank.

Similar evidence of spatial neglect can be observed in the self-portraits of Anton Räderscheidt (1892–1970) who suffered a stroke in 1967. Räderscheidt, an esteemed member of the German New Objectivity school of painting, had produced a large corpus of work prior to his stroke, with

FIGURE 7.3 Left: Self-portrait by Räderscheidt soon after his stroke showing substantial left unilateral neglect. Right: Later self-portrait showing gradual recovery of function (copyright © 2016 Petcu, Sherwood, Popa-Wagner, Buga, Aceti and Miroiu, *Frontiers in Neurology*, 7(76), pp. 1–12.

many featuring his trademark stiffly-posed couple, the male often with a bowler hat. His stroke left him with a severe left-sided neglect, as can be seen in his self-portrait from shortly after his injury (Figure 7.3, left); the left side of the canvas is completely blank, but further, the left side of the artist's face is also missing. This, combined with the additional disturbances in vision and spatial orientation, reveal much about how his stroke affected the cognitive functions associated with his damaged right hemisphere.

Over the subsequent two years of his recovery, Räderscheidt continued to paint, resulting in over 60 self-portraits from 1967 to 1969. These paintings reveal the dramatic degree of recovery that occurred over this two-year period. Figure 7.3 (right) shows the gradual return of the ability to perceive the left side of space, as this portion of the canvas is steadily filled with colours and details (though never to the same extent as the right side of the background or the face). Such recovery did not occur spontaneously, but came about through perseverance and great mental strength; as Räderscheidt himself remarked:

> Using all of my willpower, I intended to force my eyes to see correctly again.
>
> *Herzog (1991)*

In parallel with the recovery of his functionality, the post-stroke paintings of Räderscheidt seemed to have undergone a complete change in style. Although many of his earlier works portrayed couples and people realistically, they often lacked emotion and personal intimacy. Further, Räderscheidt's 'magical realist' work of the 1920s is often characterised by the use of metallic grey, blue and black colours. However, his post-stroke work used brighter colours that showed deformed couples and nudes often in wild embrace. This new colour explosion even led the artist himself to wonder how he had been able to draw without colour previously. Other cases where brain injury led to a change or enhancement of artistic style are considered next.

Art enhanced or changed by brain injury

One such change of style is evident in the work of Lovis Corinth (1858–1925), whose stroke in late 1911 seemed like the catalyst for his change from an early Impressionistic style (see Figure 7.4, left) to a later period of Expressionism (Figure 7.4, right). While it is difficult to say with certainty that this change resulted directly from his stroke, many art critics

FIGURE 7.4 Left: Example of Corinth's pre-stroke Impressionistic work, *Selbstbildnis mit schwarzem Hut und Stock* (Self-portrait with black hat and cane), Kunstmuseum 1911, St. Gallen, Switzerland. Right: Post-stroke Expressionism, *Grosses Selbstportrait vor dem Walchensee* (Large self-portrait in front of Walchensee) 1924, Bavarian State Painting Collections, Munich, Germany.

agree that his style change can be dated to early 1912. In the cases of both Corinth and Räderscheidt, the clear visual–spatial deficits caused by their brain damage were overcome to produce works of great beauty and complexity. This post-injury work can be considered of equal or perhaps even greater importance to that produced prior to their strokes. The artistic output of a recovering brain reveals much about the mechanisms of such repair.

An equally dramatic change of style was evident in the work of a Polish artist (referred to as WW) in a case report by Pachalska and colleagues. Having established himself as an artist after a difficult childhood punctuated by delinquent behaviour, trouble with the police and a diagnosis of psychosis with visual delusions, WW suffered a closed head injury in 1989 due to a traffic accident sustained while hallucinating. CT scans revealed that the accident resulted in damage to left frontal areas (atrophy), the right temporal region and the cerebellum, while cognitive testing showed the presence of left hemineglect, perseveration (an inability to break off from executing a motor behaviour) and deficits in visual memory. Most dramatic, though, was the change in WW's painting style. Prior to his accident, his works were vibrant, dream-like depictions of his hallucinatory experiences; his output during the course of his recovery and art therapy classes show many tell-tale signs of his cognitive symptoms including neglected left sides of drawings, perseverated images (often faces), and a generally 'less bizarre' content, though the images are still somewhat hallucinatory in subject matter.

While not necessarily manifested in a change of style, there have also been a number of reports where established artists have shown enhanced abilities following brain trauma. Seeley and colleagues report the case of Anne Adams (1940–present), an artist who suffered from the degenerative brain disease Primary Progressive Aphasia (PPA), a disorder that primarily affects language abilities. Although previously interested in art as a hobby, Adams had an explosion of artistic creativity over a 6-year period just before the emergence of her language deficits. Her paintings were colourful and vibrant, and culminated in a striking visual representation of Ravel's musical work *Boléro*. In *Unravelling Boléro* (Figure 7.5, top left), Adams translates the musical score into an array of vertical figures laid out in rows. The height of these figures corresponds to the volume, with a colour change marking the dramatic conclusion of the musical composition[5]. At the age of 58, two years before her language deficits were first noted, she began to paint more abstract concepts including a piece called *Pi* (Figure 7.5, top right), a visual representation of the number π using different colours to represent the first

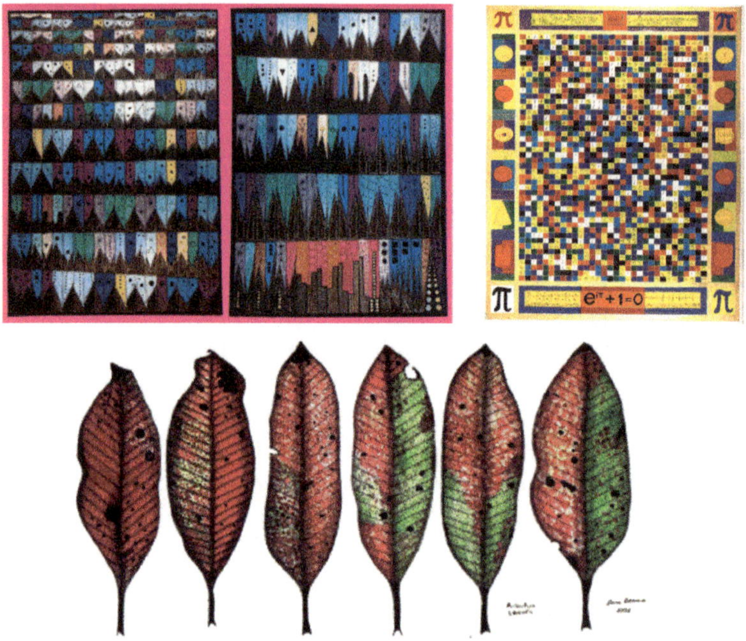

FIGURE 7.5 Examples of Anne Adams' interpretation of Ravel's musical work *Boléro* (top left), depiction of the number pi (top right), and her realistic painting of leaves (bottom). Image taken from Seeley et al. (2008), 'Unravelling Boléro: progressive aphasia, transmodal creativity and the right posterior neocortex', *Brain*, 131(Pt 1), pp. 39–49 by permission of Oxford University Press.

1,471 digits. At age 60 she began to make grammatical errors and her fluency of speech began to deteriorate. By the age of 64 she was almost mute, but despite these problems her artistic output continued. Her style, however, had demonstrably changed, her work becoming increasingly photo-realistic, often reproducing what she observed with exacting detail, realism and fidelity (Figure 7.5, bottom).

Over the course of her illness Adams underwent a number of brain scans. As expected, neuroimaging analyses revealed a severe degeneration of left frontal and temporal regions consistent with the gradual decline in language functioning; remarkably, however, her right parietal cortex showed an actual increase in grey matter volume. It was suggested that the greater neural activity in the non-dominant parietal regions may have come about due to the destruction of dominant frontal regions. As the more dominant regions began to deteriorate, regions associated with artistic creativity became liberated or disinhibited. Studies such as this not only show how the degenerating brain can affect cognitive functions, but also reveal the plastic nature of the brain and how some previously dormant or suppressed areas may take on new or enhanced roles through the complex interactions of multiple regions. Moreover, Adams' representations of music and numbers as visual shapes and colours is remarkably like the experience of certain individuals with synaesthesia, a condition discussed in the last part of this chapter.

A final, dramatic example of an enhancement of ability and change of style following brain insult is the case of patient JN, as reported by Takahata and colleagues in 2014. JN, a Japanese male who suffered a cerebral aneurysm at the age of 64, had held only a passing interest in painting after leaving school, producing a mere handful of paintings before his retirement at age 60 (two of these are shown in Figure 7.6, including a portrait of his wife painted at age 55, his last painting prior to his injury). Following his aneurysm, which involved a haemorrhagic infarction of the left prefrontal region, he showed signs of mild cognitive impairments including memory

FIGURE 7.6 Paintings by JN before (left column) and after (right four paintings) suffering from stroke. Image taken from Takahata et al. (2014), 'Emergence of realism: Enhanced visual artistry and high accuracy of visual numerosity representation after left prefrontal damage', *Neuropsychologia*, 57, pp. 38–49 with permission from Elsevier.

deficits and symptoms associated with dysexecutive syndrome (problems with planning and task execution which are common following frontal damage).

About a year into his recovery, JN began to paint again, but a dramatic shift was evident in what he produced. His post-stroke paintings were more realistic and more complex, but displayed more muted colours than his previous offerings (see Figure 7.6). In addition to the quality of his works, which mainly consisted of landscapes and portraits, he also exhibited a change in the volume of his output – he now painted for many hours each day, producing a series of works. Independent reviewers of his work from this time noted a significant shift toward realism and a significant decline in colour. A subsequent functional brain scan of JN revealed decreased perfusion in the left prefrontal cortex but – like Anne Adams – increased perfusion in the right parietal area, lending support to a growing body of evidence that the right parietal lobes may be important for drawing and painting.

Artistic ability unlocked by brain injury

The above cases all relate to the consequences of brain injury to the output of people who would be considered artists, amateur or professional, prior to their brain damage. But there are several fascinating cases where artistic abilities have emerged or are enhanced in non-artists as a result of an underlying condition. In 1996, Miller and colleagues described a patient diagnosed with frontotemporal dementia (FTD[6]) at the age of 56, who started to paint despite having no interest in art prior to this diagnosis. Over a 12-year period this patient started experimenting with colours and found himself able to draw with increasing detail and precision, to the point that he began to win prizes in local art competitions before his work gradually deteriorated. The authors report a further four cases of patients (of a larger sample of 69) with frontotemporal dementia who developed new artistic abilities; these were not merely confined to painting but extended into other domains including photography and sculpture.

While their visuospatial skills remained intact, presumably facilitating this artistic output, many of these patients showed a marked deterioration in communication and social abilities as the disease progressed. Two years later, Miller reported a further two cases of emergent artistic ability in the early stages of FTD, while Midorikawa and colleagues encountered two additional cases where patients with no history of artistic training developed realistic drawing skills after their diagnosis and during their semantic dementia.

A final example of the emergence of *de novo* artistic abilities comes from a 36-year-old female, patient MB, who suffered an acute stroke that left a weakness in her right hand. In addition, she was unable to feel warmth, heat or pinpricks on the right side of her face. She also suffered badly with burning sensations, mainly on the right upper limb (she often wore gloves to protect herself from such pain). Shortly after her stroke MB noticed a number of behavioural changes which included an increase in anxiety and a decrease in her own emotional feelings. Although she recognised emotions in others, she was unable to feel such emotions herself and often had to imitate them. This lack of emotional expression increased with time.

MB never had an interest in art prior to her stroke but within six months she started painting obsessively; she would report painting for many days, hours on end, without sleeping. Within a year her paintings were considered of a sufficient standard to be on view in local and regional exhibitions. The themes of her paintings ranged from figures and cultural scenes to abstract work and still lifes, with the majority using warm, bright colours such as red, orange and yellow (for example, Figure 7.7A). Indeed, MB reported that she felt much better as she painted, and described great pleasure and intense happiness when she used warm colours (see Figures 7.7B and 7.7C). On a number of occasions when she had to use blues, greens or colder colours (for example, Figure 7.7D), MB reported experiencing painful sensations. This is quite remarkable for a person that was unable to experience emotions. Such painful sensations only seemed to occur when she worked with the colours and not when she viewed colours as part of her own paintings or the works of others.

While there are many reports suggesting that distraction in the form of painting or other activities can serve as therapeutic relief for pain, the authors of this study suggest that this was not the case for MB. Rather, they suggest that her brain injury brought about a new perceptual experience. The stroke possibly resulted in the emergence of new associations between colour, somatosensory perception (pain) and emotion (happiness or sadness). Whether the stroke itself resulted in the liberation of multi-modal brain regions or the functional recovery post-trauma allowed other brain regions to become more active and take on extra functionality is still a matter for debate. However, MB's post-injury experience is, like that of Anne Adams, remarkably reminiscent of the cross-sensory experiences reported by some synaesthetes, as we discuss next.

Brain-based idiosyncrasies and artists

Thus far we have seen cases where damage to the brains of artists and non-artists alike has resulted in changes to their artistic outputs – deteriorations,

FIGURE 7.7 Examples of MB's first painting entitled *The Way* (A), others using warm colours (B and C), and with colder and darker colours that elicited a painful experience (for example, D). Images taken from Thomas-Anterion et al. (2010), 'De novo artistic activity following insular-SII ischemia', *Pain*, 150(1) pp. 121–7 with permission from Elsevier.

style shifts, enhancements, new skills unlocked. However, there exists another, rare, subset of those who perceive the world very differently, not due to brain injury or progressive disease, but to a quirk of genetics or a cross-wiring in the brain. Next, we will consider this population, which includes synaesthetes and tetrachromats, and the impact their conditions have on their artistic output.

Synaesthesia, literally meaning a mixing of the senses, is a genetically-based condition affecting around 4% of the population, with over 80 forms documented to date. Synaesthetes typically experience an event in one sensory modality (for example, a spoken word) in a different modality (for example, a visualised colour). In this way, a person may taste shapes, feel sounds or hear colours. The most common form, grapheme-colour synaesthesia, involves people experiencing specific colours associated with letters, numbers, days or months; for example, Mondays and the colour green. This remarkable condition appears to arise from cross-wiring in the brains of synaesthetes, such that normally disconnected sensory regions communicate more freely than in typical brains; activation of a visual area may lead to a concomitant activation of, say, an auditory region. While the existence of synaesthesia was disputed for a large portion of its history, today the condition is recognised as a genuine perceptual experience with a distinct neurological profile.

One of the more remarkable features of synaesthesia is the higher prevalence of synaesthetes who are artists. A recent survey of 358 fine arts students found that 23% of respondents reported a 'consistent' synaesthetic experience. This proportion is considerably higher than the 4% normally found in the general populace[7]. Previous research has also associated synaesthetes with higher levels of creativity, while a number of musicians, artists, scientists and writers have either self-identified as synaesthetes (e.g. Carol Steen, Mary J Blige, Pharrell Williams, David Hockney, Richard Feynman), or have been proposed as synaesthetes posthumously (e.g. Jimi Hendrix, Wassily Kandinsky, Nikola Tesla, Vladimir Nabokov). And while some synaesthetic artists endeavour not to allow their synaesthesia to manifest in their work (like Hockney), others embrace, even rely on, its presence. Artist Carol Steen experiences pain as colour, and actively attempts to represent the experience in her paintings (Figure 7.8).

FIGURE 7.8 Carol Steen's painting (oil on paper, approximately 22 × 30 inches, 1996) of what she perceives when needles are removed during acupuncture. (Image thanks to Carol Steen).

FIGURE 7.9 Wassily Kandinsky, *Composition 8*, July 1923, oil on canvas 55 1/8 × 79 1/8 inches (140 × 201 cm). Solomon R. Guggenheim Museum, New York. Solomon R. Guggenheim Founding Collection, by gift 37.262.

The same is true for many other synaesthete artists, who use their output as a means to convey and relate their unique experience of the world. Wassily Kandinsky (1866–1944), whose classification as a natural synaesthete is still debated, is credited with creating the first abstract artworks, and many of his pieces were, by his own admission, pictorial representations of musical sounds, symphonies and experiences translated into shape and colour (Figure 7.9). The following quotation certainly suggests that he was familiar with the condition:

> A certain Dresden doctor tells how one of his patients, whom he describes as 'spiritually, unusually highly developed', invariably found that a certain sauce had a 'blue' taste.
>
> *Kandinsky (1912)*

While this one strongly hints at his own experience of colour–sound relations:

> The sound of colours is so definite that it would be hard to find anyone who would express bright yellow with bass notes or dark lake with treble.
>
> *(1912)*

Whether Kandinsky's experience of synaesthesia was actual or metaphorical, his attempts to paint the music he was hearing paved the way for a new era of abstraction in art. Given that synaesthetes inhabit a world where their senses interact and comingle in complex ways, it is perhaps not surprising that so many of them are drawn to media which allow for this unique experience to be expressed, related and communicated to others.

Another, even rarer condition – tetrachromacy – again gives certain people a very different perception of their visual world, but in this case the phenomenon is restricted to one modality, that of vision. Tetrachromats, who comprise approximately 1% of the population, are capable of seeing more colours, up to 100 times more, than everyone else. Due to a rare genetic mutation, tetrachromats are born with four, rather than three, retinal cone cell types. In a normal retina, cones (specialised cells for colour vision) come in three varieties, one for each of the red, green and blue wavelengths of light. Tetrachromats possess a fourth type of cone cell, capable of processing red/orange/yellow type hues, which allows them to perceive up to 100 million colours, compared to the 1 million most people can see. Artist Concetta Antico puts her tetrachromacy to use in her paintings, imbuing her

FIGURE 7.10 *Rainbow Gully, Mission Hills* by artist Concetta Antico (with permission; see concettaantico.com).

subject matter with subtle areas of unexpected or unusual shades in order to match her visual reality (Figure 7.10). Such artworks, while stunning, serve to remind us again that the lens through which we each experience the world is determined – and often constrained – by the limitations of our physiology.

As with any exponent of a creative endeavour, the artist produces what they perceive, what they experience, what they know to be true. Visual art – whether daubs on an ancient cave wall, a self-portrait or an abstract composition – offers an insight, a snapshot into the mental, emotional or neurological state of the artist. While many brain-based conditions can bring about a stark deterioration in artistic output, other cases reported here show encouraging evidence of the potential for recovery, for enhanced ability or a change in artistic style and, even more remarkable, the emergence of artistic abilities that were not present prior to the injury. Others, by virtue of their unique physiology, perceive the world differently to most, leading them to produce enthralling works which convey their singular experience. Such individuals allow neuroscientists to examine the particular brain areas involved in these transformations, aiding our understanding of how such neurological patterns can lead to the cognitive and behavioural changes we have described. In the next chapter, we take this idea further, outlining the newly-emerging field of neuroaesthetics, the systematic exploration of the brain processes and structures responsible for the production and appreciation of art.

Notes

1 In *The Man Who Mistook His Wife for a Hat*, Oliver Sacks (2009) relates a comparable story where twelfth-century Benedictine abbess, Hildegard von Bingen, experiencing a similar phenomenon, interpreted and documented (albeit in written rather than visual form) the experience as the fall of angels from heaven.
2 See Chatterjee (2011), and Chatterjee and Vartanian (2014).
3 Forsythe, Williams and Reilly (2017).
4 Federico Fellini, the famous Italian film director, also suffered right-hemisphere stroke resulting in hemineglect. Examples of his deficit can be seen in the humorous drawings he supplied for the scientific paper: 'Preserved insight in an artist with extrapersonal spatial neglect' by Cantagallo and Della Sala (1998).
5 In an interesting coincidence, Ravel also suffered from Primary Progressive Aphasia, although Adams was unaware of this at the time of her painting. Furthermore, she painted her *Unravelling Boléro* at nearly the same age and disease stage as Ravel when he composed *Boléro*.
6 Frontotemporal dementia is a neurodegenerative disease that predominately affects the frontal and anterior temporal lobes; this contrasts with Alzheimer's disease that typically results in degeneration of the posterior parietal and medial temporal lobe regions.
7 See Urist (2016) for more details on this survey.

References

Cage, J. 1969. *A Year From Monday: New Lectures and Writings by John Cage*. Middletown, CT: Wesleyan University Press.

Cantagallo, A., and Della Sala, S. 1998. Preserved Insight in an Artist with Extrapersonal Spatial Neglect. *Cortex*. 34(2), pp. 163–189.

Chatterjee, A. 2011. Neuroaesthetics: A Coming of Age Story. *Journal of Cognitive Neuroscience*. 23(1), pp. 53–62.

Chatterjee, A., and Vartanian, O. 2014. Neuroaesthetics. *Trends in Cognitive Sciences*. 18(7), pp. 370–375.

Forsythe, A., Williams, T., and Reilly, R.G. 2017. What Paint Can Tell Us: A Fractal Analysis of Neurological Changes in Seven Artists. *Neuropsychology*. 31(1), pp. 1–10.

Herzog, G. 1991. *Anton Raderscheidt*. Cologne: DuMont.

Kandinsky, W. 1977. *Concerning the Spiritual in Art*. New York: Dover Publications.

Sacks, O. 2009. *The Man Who Mistook His Wife for a Hat*. London: Picador.

Urist, J. 2016. Why Do So Many Artists Have Synesthesia? www.thecut.com/2016/07/why-do-so-many-artists-have-synesthesia.html. Accessed 11th March 2018.

8

NEUROAESTHETICS

The machine in the ghost

Introduction

The effort to understand and explain how the mind/brain processes art and beauty has a long history. While questions of aesthetics date back to philosophers including Baumgarten (1714–1762) and Kant (1724–1804), attempts to study our responses to art and artistic creations began with the first experimental psychologists in Germany, including Gustav Fechner (1801–1887; widely regarded as the Father of Neuroaesthetics) and Hermann von Helmholtz (1821–1894). In Chapter 1 we addressed the distinctions between art, aesthetics and beauty, highlighting the controversies and open questions surrounding these concepts and how they are best defined; in this chapter, we turn our attention to the emerging field of neuroaesthetics, the systematic attempt to map the processes underpinning the perception and appreciation of art – in particular visual aesthetics – onto the structures of the brain.

We first outline the different strands of neuroaesthetics, describing the varying approaches used to address these questions. We then point out some of the principal objections to and critiques of this approach. Finally, recent models are presented, which attempt to address some of the shortcomings and provide a unifying framework for the future of neuroaesthetics. In so doing, we look to the future of neuroaesthetic research and evaluate what – if anything – it can offer our understanding of how artistic experiences have an impact on us as individuals and as a species.

Current strands of neuroaesthetics: beauty and the brain

More so now than at any other point in the history of humanity, our brains are at the forefront of our understanding of our behaviour, our personality and our place in the environment. The exponential development of technologies that offer us a window into the activities of functioning brains has resulted in a more comprehensive knowledge of how information is taken in, processed and converted into action by our brain than could have been imagined by the early pioneers of experimental psychology in the 1870s. By applying these approaches to healthy and brain-injured groups, we have learned much about the nature of such processes as memory, attention, perception, decision-making, emotion and even consciousness. It is inevitable, then, that such experimental techniques would be directed towards the question of art, art perception and visual aesthetic experience in an effort to explain the neural structures and apparatus which govern these experiences. The quest to identify the brain systems implicated in the processing of art and aesthetics can be considered to operate along three major strands: Descriptive neuroaesthetics, Experimental neuroaesthetics and Informative Anecdotes[1]. As Chapter 7 dealt with a number of informative anecdotes and what they can reveal about art and aesthetic experience, only the two former approaches will be discussed in turn here.

Descriptive neuroaesthetics

Among the first attempts to relate the activities of the brain to aesthetic experience, pioneers of neuroaesthetics VS Ramachandran and Semir Zeki pointed out the perceptual processing principles engaged when we view or appreciate works of art. Sometimes referred to as *parallelism*, this approach rests heavily on what is known about how our visual system operates (see Chapter 6) and how these principles are reflected in or engaged by works of visual art. As this explanatory approach involves extracting patterns and principles from extant works of art, it allows for the dissection of historical as well as contemporary artworks.

Our brains are constantly filling in gaps in our visual world. It is a capacity that allows us to go about our business without noticing the presence of our visual blind spot, permits the identification of partially occluded objects and makes sense of ambiguous images. The unexpected ability of our brains to allow us to perceive not what we see but rather what we *expect to see* is referred to as its 'Top Down' influence on processing. In this way, the brain fills in blanks for us automatically, allows us to perceive continuous shapes or wholes where none are present, and sometimes leads us to misread what we think are familiar phrases. While this tendency can often lead us to make errors in our perception, it reflects a truly astonishing sophistication and economy of processing in a system which is constantly bombarded by an overwhelming array of sensory stimulation. And in some cases, it drives us to want to complete the picture, finish the sculpture and fill in the gaps; famous examples of this include the Venus de Milo, the Rondanini Pieta (Figure 8.1) and Paul Cézanne's later paintings of Mont Sainte Victoire (Figure 8.2). Could part of the appeal of these pieces of art be due to our innate tendency towards completion, towards resolution? VS Ramachandran argues precisely that, proposing that unfinished pieces, or art which requires some cognitive effort or problem solving on the part of the viewer, hold a special appeal for this very reason. It is a principle called Perceptual Problem-Solving or 'Peekaboo', and

FIGURE 8.1 Left: *Venus de Milo* (By Livioandronico, 2013 [own work] [CC BY-SA 4.0], via Wikimedia Commons). Right: *The Rondanini Pieta* by Michelangelo (by Paolo da Reggio [own work] [GFDL, CC-BY-SA-3.0 or CC BY-SA 2.5-2.0-1.0]), via Wikimedia Commons.

FIGURE 8.2 Four paintings of Mont Sainte Victoire by Paul Cézanne. Upper two images: Paul Cézanne [Public domain], via Wikimedia Commons. Bottom left: Pearlman collection. Bottom right: NGI.3300, Paul Cézanne (1839–1906), *La Montagne Sainte-Victoire from Les Lauves, near Aix-en-Provence*, 1902/1904, graphite and watercolour on white paper, 47.5 × 61.5 cm, presented, Sir Alfred Chester Beatty, 1954, National Gallery of Ireland Collection, photo © National Gallery of Ireland.

it goes some way towards explaining why a partially-clad figure often carries more allure than a naked one. Ramachandran includes this principle as part of what he calls his Universal Laws of Neuroaesthetics, which also include Peak Shift (exaggerating key features for effect), Isolation (drawing particular attention to a visual element to highlight its importance) and Grouping (similar to the Gestalt principle whereby items are organised and bound into coherent groups by our visual systems)[2]. These laws, Ramachandran points out, are all rooted in the neural mechanisms underpinning visual perception, and many of them – Grouping, Symmetry, Perceptual Problem-Solving – are intrinsically rewarding processes when accomplished, thereby engaging the brain's reward circuitry and ensuring engagement and attentional capture by the artwork being perceived.

The pioneering vision scientist Semir Zeki was among the first to direct the methods of contemporary neuroscience towards the questions of art and art perception in the 1990s, and, like Ramachandran, he proposes that artists employ particular techniques to engage the brain of the viewer, with these strategies arrived at by accidental discovery or years of trial and error. These visual tricks, Zeki claims, are effective precisely because they activate fundamental neural processes which give rise to the desired effect. He points to the key principles of Constancy – how the brain's ability to accommodate invariant properties of an object at different viewing angles is mimicked by the artist's ability to distil the essence of a visual object into a single representation – and Abstraction – how the artist's creation of abstract representations parallels the brain's ability to generate abstractions of representations due to cognitive economy or processing constraints – as core examples. He proposes that artists can be considered to be amateur scientists, experimenting and exploring the different ways in which the brain can best be engaged through the application of particular techniques[3].

A further prominent pattern emerging from work in descriptive neuroaesthetics is the important role of the Default Mode Network (DMN) in art appreciation and perception. The DMN is a network of brain regions (including lateral and medial prefrontal cortex, posterior cingulate cortex, temporo-parietal junction, superior frontal gyrus and hippocampal formation) which appears to be activated while the brain is not actively engaged in any ongoing processing task, i.e. when the brain is effectively 'idling'. The DMN has been proposed to correspond with the experience of 'disinterested interest'. This phenomenon is said to describe the type of engagement associated with viewing works of art. Such DMN activations have been shown when people rate visual artworks as 'highly moving', and are suggested to be indicative of the engagement of self-referential processes like autobiographical memory or the sense of self[4].

Experimental neuroaesthetics

In parallel with these descriptive approaches, experimental neuroaesthetics seeks to investigate the neural phenomena associated with aesthetic experience using the techniques of experimental psychology and neuroscience, often incorporating functional brain imaging. This approach has led to a focus on the processing stages and associated brain regions involved in the Aesthetic Triad – a trio of mechanisms responsible for sensory–motor, emotion–evaluative, and meaning/contextual responses to aesthetic experiences[5]. Sensory–motor areas typically involve visual processing regions engaged by the content of what is being perceived – fusiform face area (FFA) for portraits, visual area V4 for colour, for example – while motor areas may be engaged via the mirror neuron system in some cases (e.g. when viewing a dancer performing a piece). Some of these regions may also be active during the emotion–evaluative stage of processing, this phase of the artistic perceptual experience also engages brain areas specialised for emotion (for example, the amygdala) and reward (the frontal reward system network), as well as the Default Mode Network. In this vein, scientists claimed to have localised the processing of 'beauty' in the brain to a region of medial orbitofrontal cortex (mOFC), though this conclusion is contentious[6]. The third aspect of the triad, the subsequent processing of meaning or cultural context (including the reputation of the artist, the intended meaning of the piece and other considerations), is often considered neglected or deficient in many theories of neuroaesthetics due to the difficulty of operationalising it. However, it has been proposed that the study of conceptual art offers a potential solution to this problem, allowing a novel window into the study of meaning in art[7].

Such experimental approaches have yielded four main theories of experimental neuroaesthetics: Formalist, Contextual, Mimetic and Expressionist. These theories, which seek to explain how we respond to and evaluate works of art, are briefly described as follows:

1. Formalist theories propose that visual aesthetic experience relies heavily on the formal properties of the visual stimuli being perceived; these properties are considered to be universal (similar to Ramachandran's Laws, above), and therefore the emphasis is placed heavily on the presence of specific key qualities in the artwork itself, including beauty.
2. Contextual theories, by contrast, take greater account of the intention of the artist and the circumstances surrounding the creation and/or display of an artwork. In such theories, considerations of beauty play a lesser role, meaning that the response to a piece will be determined by the viewer's judgement of the contextual characteristics of the work. Such theories are particularly useful for explaining responses to postmodern or contemporary art.
3. Mimetic theories apply mainly to representational forms of art, whereby art is rated and evaluated based on how well the piece mimics the subject which it seeks to depict.
4. Expressionist theories propose that evaluations are based on how the artist manages to convey their emotional state to the perceiver via their artwork, with judgements based on the degree of success of this attempt.

The experimental approach to neuroaesthetics and the identification of the Aesthetic Triad appears to offer a formal, objective and scientific roadmap for the study of aesthetic experience; however, a number of objections and criticisms have been offered, highlighting the potential limitations and biases inherent in these approaches. It is to these criticisms that we next turn our attention.

Objections to and critiques of neuroaesthetics

Several of the key objections to neuroaesthetics concern the nature of what is actually being studied, and so are, therefore, rooted in definitions. Important distinctions can be drawn between the perception of art and visual aesthetic experience, while some have pointed out that *aesthetics is not the same thing as art appreciation*[8]. Others highlight the strong *visual* (and to a lesser extent auditory) *bias* in what is studied, citing the neglect of artforms such as prose fiction under the neuroaesthetic umbrella[9]. A *beauty bias* in the research to date has also been noted, wherein the concept of beauty is frequently (and inappropriately) conflated with that of art or aesthetic experience[10] (see Chapter 1 for a discussion of the relationship between art, aesthetics and beauty).

Commentators in the field also lament the neglect – often for practical reasons – of any consideration of *meaning and/or historical context* in neuroaesthetic investigations, emphasising the important role played by such evaluations in the aesthetic experience[11]. As mentioned above, conceptual art may offer a means to disentangle issues of meaning and objecthood from considerations of beauty and preference that often confound experiments of this nature. Meanwhile, Chatterjee points out three major challenges for neuroaesthetics to address: (1) the *risk of reduction*, whereby the simplifications and modifications needed to make a phenomenon like aesthetic experience amenable to experimental testing run the risk of losing the phenomenon of interest; (2) the distinction between the investigation of *brain activity* and the investigation of *aesthetic experiences*, which are not the same; and (3) the larger question of *what added value is provided* to our understanding of the aesthetic experience by having knowledge of the brain systems activated[12].

This last point – what added value does neuroaesthetics provide – is perhaps the most emotive argument in relation to this topic. Philip Ball, in his evocatively-titled commentary 'Neuroaesthetics is killing your soul', outlines some key theoretical problems with the endeavour to isolate brain regions associated with responses to art. Ball raises concerns about the ability of the rational reductionist approach to yield universal principles and/or the neural basis of beauty (the conflating of art and beauty is discussed in Chapter 1 of this volume). He further highlights the risks associated with drawing up a set of rules for critical judgements of art based on brain responses, suggesting (perhaps unkindly, and without any supporting data) that scientists will almost inevitably put their findings to work in this manner. His closing argument against neuroaesthetics is as follows:

> There are certain to be generalities in art and our response to it, and they can inform our artistic understanding and experience. But they will never wholly define or explain it.
>
> *Ball (2013)*

Models and frameworks for neuroaesthetics

Despite these objections – and in some cases in response to them – several attempts have been made to formulate general, overarching and unifying models of neuroaesthetics to facilitate the future trajectory of the field. Three of these models are briefly summarised below.

Marin's General Model (2015): experimental psychologist Manuela M Marin proposes a wider hinterland for neuroaesthetics, moving beyond the bias towards visual aesthetic experience to encompass both the 'sister arts' of painting, poetry and music, and other artforms such as literature and dance. Marin claims that this will require a broader definition of aesthetics, leading to a General Model of Neuroaesthetics which can encompass other senses beyond the solely visual, with the flexibility to consider different object classes within each sensory modality and to accommodate cross-domain or multisensory aesthetic experiences.

Redies' Unifying Model (2015): neuroscientist Christoph Redies attempts to combine universal beauty and cultural context within a unifying model of visual aesthetic experience. Two parallel processing modes are proposed, one perceptual and one cognitive. The perceptual arm of the model is largely driven by 'Bottom Up' (or sensory-driven) processes and deals with the intrinsic form of the artwork and universal perceptual responses to it, leading to an evaluation of the aesthetics of perception for that piece. The cognitive arm, which is partially Top Down in nature, varies across individuals and takes account of content, context and cultural experiences associated with the work, resulting in an aesthetics of cognition evaluation. The model proposes that, if both evaluations are favourable, this will normally result in an overall positive visual aesthetic experience. However, exceptions to this are cited, where only one positive evaluation is sufficient to elicit a favourable overall response. Examples include postmodern/contemporary art. Redies also emphasises the role of emotional processing, again citing a possible role for the brain's Default Mode Network in the state of 'disinterested interest/pleasure' evoked by pleasing works of visual art.

Bullot and Reber's Psycho-Historical Framework (2013): proposed by philosopher Nicolas J Bullot and psychologist Rolf Reber, this model critiques the lack of emphasis on historical context or the intention of the artist within neuroaesthetics. The authors claim that context and intention are often critical aspects of artistic works, underscoring the difference between aesthetic experience and art appreciation. Namely, 'appreciators of art' are typically aware of, or sensitive to, the art-historical context of artworks, giving them an insight into the functions and intentions of the work. In contrast, variations in the degree of this sensitivity are ignored by experimental neuroaesthetic studies where brain responses are combined and averaged across multiple participants. The proposed Psycho-Historical Framework considers the psychological responses to, and the causal/historical information carried by, artworks, wherein the intention of the artist and the knowledge level (or sensitivity) of the appreciator are taken into account. Bullot and Reber identify three modes of art appreciation, Basic Exposure – an elementary mode couched principally in perceptual processes – the Artistic Design Stance – a deeper level of appreciation involving some knowledge of art-historical context and intended impact – and finally Artistic Understanding, the deepest level of art appreciation wherein the appreciator can explain the artistic status and/or function of the work. The authors claim that neuroaesthetics to date has concerned itself almost exclusively with the Basic Exposure level of aesthetic experience, and state that any complete science of art and aesthetic experience must also take account of the two deeper levels.

Conclusion

Few emerging areas within neuroscience in recent years have proved as controversial as that of neuroaesthetics. For many, the proposition that we can now begin to explain the governing principles and anatomical mechanisms that allow us to appreciate a work of art/visual aesthetics as beautiful is to rob that piece of its mystery, its essence or even its beauty itself. But neuroaesthetics purports to elucidate *how* the brain appreciates art, not *why* it appreciates it. There is nothing inherent in that goal which should strip the art of any of its mystique. On the contrary, understanding how something captures our attention, stimulates our interest and elevates our spirit is to draw back the veil and allow us to appreciate its beauty on a different, deeper level; it may even give us access to a vocabulary previously inaccessible that allows us to express more clearly the effect that art has on us. It is difficult to see how that can be a bad thing.

The claim that the brain is at least partially hardwired to appreciate art is sometimes used as a criticism of neuroaesthetics, that such assertions highlight the hubristic folly of the reductionist approach to such a topic. But this is, in many ways, a straw man argument – it may be more correct to state that art is, in fact, at least partially designed to be appreciated by the brain. Walk around any gallery in the world once you have been made aware of Ramachandran's Laws and you will find your experience forever changed. For once you are made aware of them, you will see their signatures in every piece you view – an echo of grouping and symmetry here, a hint of peak shift there, the stark beauty of isolation, the tantalising allure of the 'peekaboo effect'. In essence, once you are aware of these laws, you cannot then unsee them in each new work you encounter. And far from diminishing their beauty or taking from the wonder of their creation, an awareness of these laws should help us to marvel with even greater awe at the ingenuity of the artist in deriving such rules organically, to appreciate even more the achievement of such (sometimes) amateur scientists in stumbling across such laws through years of experience, trial and error, failure and eventual triumph. It should amplify the accomplishment of such artists, not diminish it.

The concerns raised by Philip Ball and others regarding how neuroaesthetics threatens to explain away artistic/aesthetic experience in purely neural terms, and to strip these processes of their magic and mystery, are unfounded. On the one hand, many of these arguments take aim squarely at straw men, bemoaning the risks inherent in what they claim neuroaesthetics purports to do, rather than the actual intention of this emerging discipline. The attempt to elucidate neural processes that govern our responses to art and aesthetics is, by definition, one of discovery, of insight and understanding. By furthering our knowledge of what happens in our brains when we view, appreciate and react to a work, the net result can only be to add to, not take away from, the act of experiencing it. The neuroscientific study of memory, our enhanced understanding of the processes underpinning encoding and subsequent retrieval of events in our brains, has done nothing to dilute the nostalgic feelings and emotions evoked by recollecting a treasured childhood experience; this is true of neuroscientists as much as it is for the public. Why would it be different for art?

To consider the imaginative and clever ways in which artists have, since the beginnings of civilisation, pushed at the limits of these laws, tested their boundaries and established their thresholds is an enormously enriching exercise. It may be true to state that, when an artist feels that a piece 'needs' a splash of colour in a particular corner, it is in fact their *brain* that needs it there – does that make the process, the endeavour, the creativity any less human? Or more human? Does understanding how a magic trick works take from its effect, or add to one's appreciation of its ingenuity? Perhaps it is the same with neuroaesthetics – it is inevitable that it will be a polarising topic. In this case, it may be best to effectively agree to disagree; for those who do not enjoy seeing how a magic trick works, perhaps leave neuroaesthetics to those who do.

In Chapter 1 we discussed the views of those who argue that art has contributed virtually nothing to science, and that the two disciplines should cease the pretence that they have any similarities worth vaunting. We disagree. In the early chapters of this volume we provided many examples of how (sometimes aesthetically pleasing) visual representations have been pivotal in the history of science and its development in many fields of study, from the Renaissance anatomists to Galileo and Cajal and the pioneers of data visualisation such as von Humboldt and Nightingale. Modern science continues to be informed and enriched by novel data visualisation approaches which remain informed by many principles and influences from visual aesthetics.

In the later chapters, we highlighted the myriad ways in which, often unintentionally, artworks have shed light on scientific phenomena, most commonly (and appropriately) in relation to the functioning of our brains. While critics are of course correct to point out that scientists and artists have different goals, different agendas and a separate repertoire of tools and systems with which to pursue their interests, the contention that the two groups share no similarities and can offer little to each other – or, more specifically, that art can offer little to science – seems pompous. It would perhaps be more accurate to suggest the two are *symbiotic* rather than similar. The examples given here, both historical and modern, show how many of the eminent scientific minds of their respective eras relied on artistic considerations, if not for their investigations themselves then at least for their dissemination. Further, Ramachandran's and Zeki's laws of neuroaesthetics demonstrate how artists have, over the centuries, acted almost as amateur scientists, engaged in their own exercise of discovery within their own medium. Art and science have been, and continue to be, closer bedfellows than opponents appear to realise.

Critics go on to emphasise the fundamental difference between the approaches, that science has only one correct answer while art is open to multiple personal interpretations. They further state that, for this reason, art can be appreciated by anyone, while to make a meaningful contribution to a scientific debate or endeavour, one must possess a detailed knowledge of the subject matter. While this is again true, critics miss the opportunity presented by this state of affairs, one which would be lost should their desire to segregate the fields be realised. Namely, because art can be appreciated by all, it can provide a means to those without detailed knowledge of a concept or phenomenon to glimpse, even briefly, some of the wonder, mystery, splendour and even *beauty* that science can unveil. In this way, initiatives like 'The Art of Science'[13], 'Art the Science'[14] and 'Interstellate'[15], and agencies which fund such collaborations, including the UK's Wellcome Trust, contribute to the wider accessibility of scientific findings. This enables the public to experience, even to a limited degree, a flavour of the enchantment which drew many of us into science in the first place. To preclude this possibility, to denigrate its value, as some scientists apparently desire to do, strikes us as foolish.

The long and complex relationship between art and science has wound its way through the centuries on an erratic and meandering path, initially intertwined and closely joined, later diverging onto parallel routes. But for all the many changes in trajectory, the stories related here suggest that art has played a key role in many scientific advances, sometimes in bizarre or unexpected ways, and science in turn has had a significant impact on art. At a time when the investigation of the brain's response to artistic output is yet in its infancy, heralding a possible convergence of the two once more, it is inevitable that there will be voices of objection from both camps. It is our belief that, as history has shown, such intermingling of art and science is likely to enrich both, not because they are similar, but precisely because they are different. Exposure to such diversity of thought and methods can surely be a positive thing, if only for the simple function of initiating the discussion and prompting efforts to derive a common language in which the contributors might converse. Neither art nor science has anything to fear from the other; there are only things to be gained. In *The Demon-Haunted World*, Carl Sagan wrote:

> There are no forbidden questions in science, no matters too sensitive or delicate to be probed, no sacred truths.
>
> *Sagan (2011, p. 34)*

Why, then, should art, artistic experience, aesthetics and such associated concepts evoke such resistance from artists and scientists alike? Our comprehension of how a piece of art moves us, stirs us, captures our attention and evokes sometimes complex and contradictory responses in us is a challenge to explain some of the most fundamental aspects of what makes us human. We believe that this challenge is one that science should embrace and exalt in, not shy away from. And while doubtlessly fraught with difficulties and obstacles, any meaningful progress into our understanding of how art affects us will only be accomplished in collaboration with artists, working closely with them, learning from them. The success of this venture will depend on such inclusivity.

Notes

1 Chatterjee (2011), and Chatterjee and Vartanian (2014).
2 For a more comprehensive description of these laws, see Ramachandran (2010).
3 See Zeki (1999) for a more detailed explanation.
4 See Vessel, Starr and Rubin (2013) for more explanation.
5 See Chatterjee and Vartanian (2014).
6 See Kawabata and Zeki (2004).
7 See Kranjec (2015).
8 Bullot and Reber (2013).
9 Burke (2015).
10 Pepperell (2015) and Consoli (2015).
11 Kranjec (2015), Siler (2015) and Bullot and Reber (2013).
12 Chatterjee (2011).
13 A virtual art museum inspired by the world of science, curated by Dr Henrietta Bowden-Jones.
14 A Canadian science–art non-profit organisation.
15 An art–science outreach initiative curated by Caitlin M Vander Weele.

References

Ball, P. 2013. Neuroaesthetics is Killing Your Soul. *Nature* (22 March 2013). www.nature.com/news/neuroaesthetics-is-killing-your-soul-1.12640.

Bullot, N.J., and Reber, R. 2013. The Artful Mind Meets Art History: Toward a Psycho-historical Framework for the Science of Art Appreciation. *The Behavioral and Brain Sciences*. 36(2), pp. 123–137.

Burke, M. 2015. The Neuroaesthetics of Prose Fiction: Pitfalls, Parameters and Prospects. *Frontiers in Human Neuroscience*. 9(442), pp. 1–12.

Cantagallo, A., and Della Sala, S. 1998. Preserved Insight in an Artist with Extrapersonal Spatial Neglect. *Cortex*. 34(2), pp. 163–189.

Chatterjee, A. 2011. Neuroaesthetics: A Coming of Age Story. *Journal of Cognitive Neuroscience*. 23(1), pp. 53–62.

Chatterjee, A., and Vartanian, O. 2014. Neuroaesthetics. *Trends in Cognitive Sciences*. 18(7), pp. 370–375.

Consoli, G. 2015. From Beauty to Knowledge: A New Frame for the Neuropsychological Approach to Aesthetics. *Frontiers in Human Neuroscience*. 9(290), pp. 1–3.

Kawabata, H., and Zeki, S. 2004. Neural Correlates of Beauty. *Journal of Neurophysiology*. 91(4), pp. 1699–1705.

Kranjec, A. 2015. Conceptual Art Made Simple for Neuroaesthetics. *Frontiers in Human Neuroscience*. 9(267), pp. 1–5.

Pepperell, R. 2015. Artworks as Dichotomous Objects: Implications for the Scientific Study of Aesthetic Experience. *Frontiers in Human Neuroscience*. 9(295), pp. 1–13.

Ramachandran, V.S. 2010. *The Tell-Tale Brain*. London: W.W. Norton & Company.

Sagan, C. 2011. *The Demon-haunted World: Science as a Candle in the Dark*. New York: Ballantine Books.

Siler, T. 2015. Neuroart: Picturing the Neuroscience of Intentional Actions in Art and Science. *Frontiers in Human Neuroscience*. 9, pp. 1–8.

Vessel, E.A., Starr, G.G., and Rubin, N. 2013. Art Reaches Within: Aesthetic Experience, the Self and the Default Mode Network. *Frontiers in Neuroscience*. 7(258), pp. 1–9.

Zeki, S. 1999. *Inner Vision: An Exploration of Art and the Brain*. Oxford: Oxford University Press.

INDEX

References to figures are shown in *italics*. References to footnotes consist of the page number followed by the letter 'n' followed by the number of the note, e.g. 46n4.